GRAPHING CALC[illegible]
EXCEL® SPREADSHEET MANUAL

FINITE MATHEMATICS AND CALCULUS WITH APPLICATIONS SERIES

EIGHTH EDITION

Margaret L. Lial
American River College

Raymond N. Greenwell
Hofstra University

Nathan P. Ritchey
Youngstown State University

Boston San Francisco New York
London Toronto Sydney Tokyo Singapore Madrid
Mexico City Munich Paris Cape Town Hong Kong Montreal

Contributors:

Bill Ardis
Collin County Community College

Raymond N. Greenwell
Hofstra University

Thomas Hungerford
Saint Louis University

Todd Lee
Elon College

Thomas W. Polaski
Winthrop University

Nathan P. Ritchey
Youngstown State University

Paula Grafton Young
Salem College

Alan Ziv

Excel is a registered trademark of Microsoft Corporation.
Reproduced by Pearson Addison-Wesley from QuarkXPress® files.

Publishing as Pearson Addison-Wesley, 75 Arlington Street, Boston, MA 02116.

ISBN-13: 978-0-321-45067-8
ISBN-10: 0-321-45067-1

2 3 4 5 6 BB 10 09 08

CONTENTS

GRAPHING CALCULATORS

EXCEL SPREADSHEETS

Part I

General Instructions for Graphing Calculators

Graphing Calculators

PART 1 INTRODUCTION

Graphing calculators are generally superior to ordinary scientific calculators, not only for graphing (which you would expect), but also for most multi-step calculations. Their screens typically display 8 lines of text, so you can easily see what numbers and functions have been entered. You can easily edit or change these entries, without losing your place in the computation.

Although there are graphing calculators on the market for as little as $40, you would be well advised to use care when buying the less expensive ones, many of which lack useful features that are standard on more expensive models. Before buying a graphing calculator, you should consider which features you are most likely to need. The following chart may be helpful. Many of the features mentioned in the table are discussed in the next part of this supplement.

Feature	Used In FM	Used in CWA	Used in FM and CWA
Table Making Capability	Chapter 1	Chapters 1, 2, 3, 7	Chapters 1,
Root Finder (Equation Solver)		Chapter 6	Chapter 14
Financial Functions	Chapter 5		Chapter 5
Matrix Operations	Chapters 2, 4, 10, 11		Chapters 2, 4,
System of Linear Equations Solver	Chapters 2, 10, 11		Chapters 2, 4
Statistical Operations	Chapters 1, 9	Chapters 1, 2, 11, 13	Chapters 1, 9, 10, 18
Maximum/Minimum Finder		Chapters 5, 6	Chapters 13, 14
Intersection Finder	Chapter 1	Chapters 1, 2	Chapters 1, 10
Numerical Derivatives		Chapters 3, 4	Chapters 11, 12
Numerical Integration		Chapter 7	Chapter 15

Current models that have most or all of these features include TI-82/83/83 Plus/84 Plus/85/86/89/92. Most graphing calculators have additional features that do not play a role in this book (such as polar coordinates, parametric graphing, and complex numbers). A few, such as the TI-89 and TI-92, can perform symbolic operations such as factoring and finding derivatives).

ADVICE ON USING A GRAPHING CALCULATOR

1. **BASICS** Graphing calculators have forty-nine or more keys. Most modern desk-top computers have 101 keys on their keyboards. With fewer keys, each key must be used for more actions, so you will find special mode-changing keys such as "**2nd**", "**shift**", "**alpha**", and "**mode**". Become familiar with the capabilities of the machine, the layout of the keyboard, how to adjust the screen contrast, and so on.

2. **EDITING** When keying in expressions, you can pause at any time and use the **arrow keys**, located at the upper right of the keyboard, to move the cursor to any point in the

text. You can then make changes by using the "**DEL**" key to delete and the "**INS**" key to insert material. ("INS" is the "second function" of "DEL.") After an expression has been entered or a calculation made, it can still be edited by using the "**ENTRY**" feature. On TI calculators, use "2nd, ENTER" to return to the previously entered expression.

3. **SCIENTIFIC NOTATION** Learn how to enter and read data in **scientific notation** form. This form is used when the numbers become too large or too small (too many zeros between the decimal point and the first significant digit) for the machine's display.

4. **FUNCTION GRAPHING**

   A. **Function Memory** Learn how to enter functions in the "**function memory**" (labeled "Y="), and to mark them to be graphed. Depending on the calculator, you can store from 10 to 99 functions in the function memory, so that you don't need to key them in each time you want to use them.

   B. **Viewing Window** Use the "**RANGE**" key (labeled in "**WINDOW**" on some calculators) to determine what part of the coordinate plane will appear on the screen. You must key in the minimum and maximum values for x and y as well as the distance between the tick marks on the axes (for instance, Xscl = 1 and Yscl = 2 means that tick marks will be one unit apart on the x-axis and 2 units apart on the y-axis).

   C. **Graphing Mode** Normally your calculator is set to graph in "**connected mode,**" meaning that it plots the points and connects all of them with a continuous curve (essentially what you usually do when graphing by hand, except that the calculator plots more points). Sometimes (particularly when there should be breaks in the graph, such as a vertical asymptote) connected mode graphing produces misleading or inaccurate graphs. On these occasions, you may want to change the graphing mode to "**DOT,**" so that the calculator will plot the points, but won't connect them.

   D. **Trace** With the "**Trace**" feature, the left/right arrow keys can be used to move the cursor along the last curve plotted, and the values of x and y will be displayed for each point plotted on the screen. On all calculators, if more than one graph was plotted, you can move the cursor vertically between the different graphs by using the up/down arrow keys.

   E. **Zoom** The "**zoom**" feature allows a quick redrawing of your graph using smaller ranges of values for x and y ("**zoom in**") or larger ranges of values ("**zoom out**"). Thus one can easily examine the behavior of a function within the close vicinity of a particular point of the general behavior as seen from far away.

   F. **Plot** With "**Pt-On**" you can move the cursor to any point on the screen and either have the machine plot a single point or have it display the "screen coordinates" of the point. For example, on the TI, the command "Pt-On(2,3)" will cause this point to be plotted on the graph screen. However, the "Screen coordinates" will usually be approximations of the specified values.

SOLVING EQUATIONS

1. **EQUATIONS OF THE FORM** $f(x)=0$. A graphing calculator produces highly accurate approximations of the solutions of equations such as

$$x^3-2x^2+x-1=0.$$

To solve this equation, graph the function $x^3-2x^2+x+1=0$ (see Figure 1). The places where the graph touches the x-axis (the x-intercepts of $y=f(x)$) are the values of x that make $f(x)=0$, that is, the solutions of the equation. Figure 1 suggests that in this case, there is just one x-intercept, located between $x=1$ and $x=2$. On a typical calculator it can be precisely located in several ways.

Graphical Root Finder In the CALC or MATH menu, look for a key labeled "ZERO" or "ROOT." The syntax for this command varies with calculator. On some calculators, you will be asked to specify a lower and upper bound (that is, numbers on either side of the x-intercept you are seeking) and possibly to make an initial guess. Check your instruction manual for details. The root finder on a TI-83/84 Plus shows that the solution is $x\approx 1.7548777$ (Figure 2).

Zoom-in Use the zoom-in feature of the calculator, or repeatedly change the range settings by hand, so that only a very tiny portion of the graph near the x-intercept is shown. Figure 3, in which the endpoints of the x-axis are 1.7 and 1.8, shows the final window in such a process. The tick marks on the x-axis are .01 unit apart and the desired x-intercept is between 1.75 and 1.76, at approximately 1.754 or 1.755. This is very similar to the approximation obtained by the Root Finder.

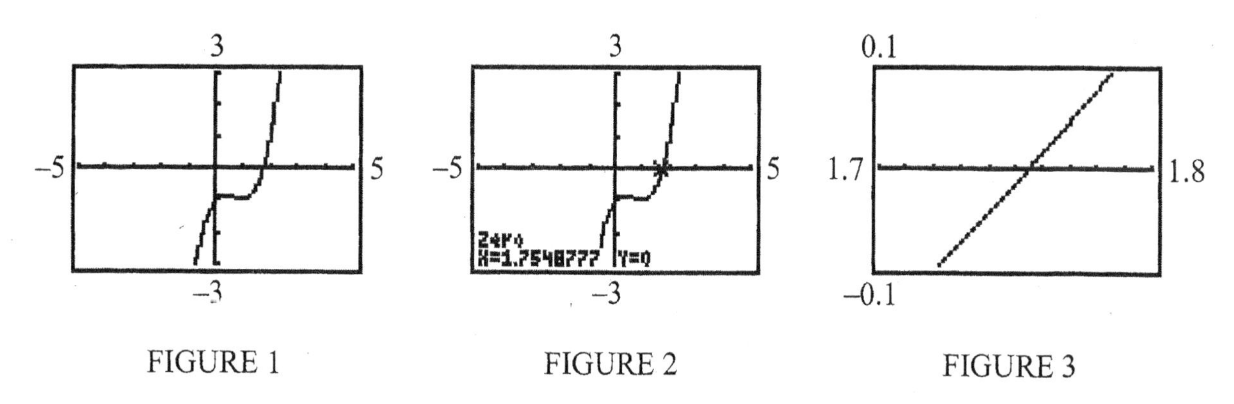

FIGURE 1 FIGURE 2 FIGURE 3

Algebraic Equation Solver Some calculators have an algebraic equation solver as well, which will solve an equation without the need to graph it first. Check your instruction manual for the correct syntax.

2. **SYSTEMS OF TWO EQUATIONS IN TWO UNKNOWNS** Graphing calculators can solve systems of equations in two variables by finding the intersection points. For example, to solve the system

$$y=x^3-2x^2+x-3$$
$$y=4x^2-3x-7,$$

graph both equations on the same screen, as in Figure 4. The points where the graphs intersect are the points (x, y) that satisfy both equations, that is, the solutions of the system. They can be found in two ways.

Graphical Intersection Finder In the CALC or MATH menu, look for the key labeled "INTERSECTION" or "ISECT." The syntax for this command varies with the calculator. On some calculators, you will be asked to specify the two curves and possibly to make an initial guess. Check your instruction manual for details. The intersection finder on a TI-83/84 Plus (Figure 5) shows that the intersection point to the right of the y-axis has coordinates

$$x \approx 1.482696 \text{ and } y \approx -2.654539.$$

This is one solution of the system.

Zoom-in Use the zoom-in feature of the calculator, or repeatedly change the range settings by hand, so that only a very tiny portion of the graph near one of the intersections points is shown. Figure 6 (graphed in "grid on" mode) shows the final window in such a process. The grid is determined by the tick marks on the axes, which are 0.01 unit apart. Using the trace feature we estimate that the intersection point has coordinates $(-0.53334, -4.257)$. So the other solution of the system is $x \approx -0.534$ and $y \approx -4.257$.

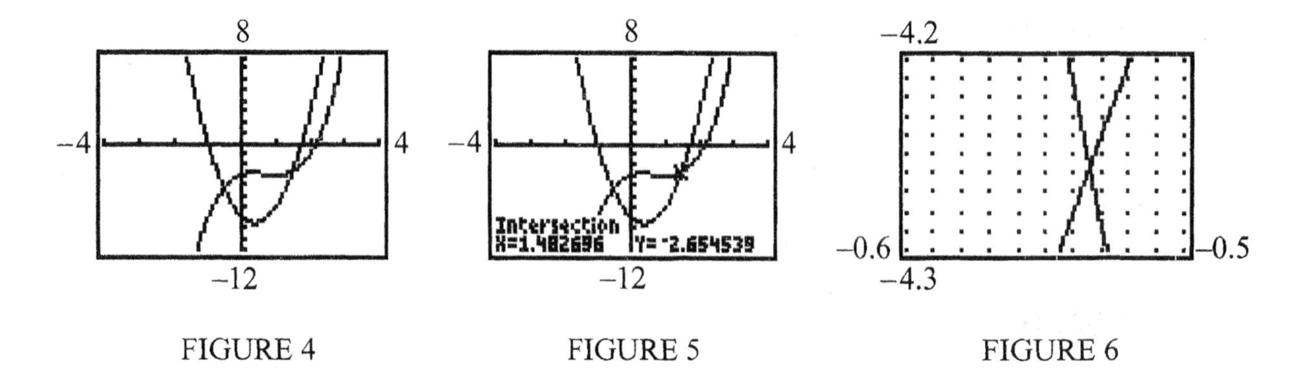

FIGURE 4 FIGURE 5 FIGURE 6

3. **SYSTEMS OF LINEAR EQUATIONS** Systems of two linear equations in two unknowns can be solved by the methods of the preceding paragraphs. Some calculators have a built-in solver for systems of linear equations and all graphing calculators have matrix capabilities that enable you to solve larger systems by using row operations or matrix inverses. These techniques are described in Chapter 2 of *Finite Mathematics*.

FINDING MAXIMUM AND MINIMUM FUNCTION VALUES

The graph of $y = x^3 + x^2 - 3x - 2$ is shown in Figure 7. A graphing calculator can find the local maximum and minimum values of this function (which correspond graphically to the "top of the hill" to the left of the y-axis and the "bottom of the hill" to the right of the y-axis). As with roots and intersections, this can be done in several ways.

Maximum/Minimum Finder To find the local maximum (top of the hill), look in the CALC or MATH menu for a key labeled "MAXIMUM" or "MAX." The syntax for this command varies with the calculator. On some calculators, you will be asked to specify upper and lower bounds and possibly to make an initial guess. Check your instruction manual for details. The maximum finder on a TI-83/84 Plus (Figure 8) shows that the local maximum occurs when $x \approx -1.3874$ and $f(x) \approx 1.4165$.

Zoom-in Use the zoom-in feature of the calculator, or repeatedly change the range settings by hand, so that only a very tiny portion of the graph near the local minimum point is shown. Figure 9 (in "grid on" mode) shows the final window in such a process. The grid is determined by the tick marks on the axes, which are 0.01 unit apart on the x-axis and 0.001 unit apart on the y-axis. Using the trace feature we estimate that the local minimum point has approximate coordinates $(0.721, -3.268)$. So the local minimum occurs when $x \approx 0.721$ and $f(x) \approx -3.268$.

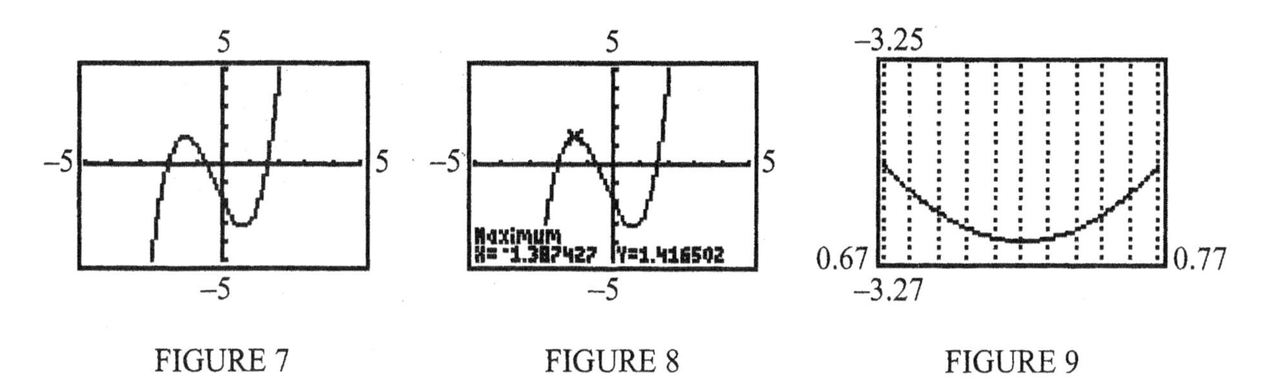

FIGURE 7 FIGURE 8 FIGURE 9

PROGRAMMING

Many relatively complicated formulas and processes must be used frequently (for example, the various financial formulas in Chapter 5 and the simplex method in Chapter 7). In such cases you can use the programming feature of a graphing calculator to automate the process. Once a program has been entered into the calculator, it can be called up whenever needed. Typically, the program asks the user to enter the necessary data (for instance, the amount of the loan, the interest rate, and the number of payments) and then quickly produces the desired information (for example, the monthly payment for the loan). Programming syntax and procedures vary from calculator to calculator, so check your instruction manual.

The textbooks website (www.aw.com/LGR) contains useful programs for many graphing calculators. These programs are of two types: programs to give older calculators some of the features that are built-in to newer ones (such as a table maker for TI-85) and programs to do specific tasks (such as financial formulas, amortization tables, and the simplex method).

SOME SUGGESTIONS FOR REDUCING FRUSTRATION

We all find ways to make even the simplest machines do the wrong things without even trying. On of the more common problems with graphing calculators is getting a blank screen when a graph was expected. This usually results from not setting the "**Range**" values appropriately before graphing the function, although it could also easily result from incorrectly entering the function.

Another common error is having more right parentheses than left parentheses. To confuse us further, these same calculators think it is perfectly OK to have more left parentheses than right parentheses. For instance, the expression $5(3-4(2+7)$ has two left parentheses and one right parenthesis, but it will be evaluated as $5(3-4(2+7))$.

The message "ERR" or "ERROR" appears when a number is too large or when a number is not allowed. If you try to find the 1000th power of ten or divide a number by zero you will most certainly see some kind of error message. When the TI models detect an error, they display a special menu that lists a code number and a name for the type of error. For certain types of errors the choice "**GOTO**" is offered. The TI-86 will display the number 9.99999999E999 but shows "ERROR 01 OVERFLOW" for 10E999 (which means 10 times the 999th power of ten). The latter and other more advanced calculators display "(0, 2)" when asked to find the square root of -4. The "(0, 2)" represents the complex number $0+2i$.

When graphing **rational functions** you sometimes will see strange little "blips" in an otherwise smooth graph. This usually means there is a **vertical asymptote** at that location due to division by zero, but you cannot see the actual behavior there because your "**window**" is too large. "**Zoom in**" on the graph in the region of the irregularity to obtain a better view. Also, as mentioned earlier in this section, one can switch to **dot mode** to eliminate unwanted lines in the graph.

SOME FINAL COMMENTS

While studying mathematics it is important to learn the mathematical concepts well enough to make intelligent decisions about when to use and when not to use "high tech" aids such as computers and graphing calculators. These machines make it easy to experiment with graphs of mathematical relations. One can learn much about the behavior of different types of functions by playing "what if" games with the formulas. However, in a timed test situation you may find yourself spending too much time working with the graphing calculator when a quick algebraic solution and a rough sketch with pencil and paper are more appropriate.

To get the most return on your investment, learn to use as many features of your calculator as possible. Of course, some of the features may not be of use to you, so feel free to ignore them. A first session of two or three hours with your graphing calculator and your user's manual is essential. Be sure to keep your manual handy, referring to it when needed.

A final word of caution: These calculators are fun to use, but they can be addictive. So set time limits for yourself, or you may find that your graphing calculator has been more of a detriment than a help!

Part II

Detailed Instructions for Graphing Calculators

Detailed Instructions for Finite Mathematics

This section of Part II contains detailed instructions for using the TI-83 and 83/84 Plus, TI-85, TI-86 and TI-89 for *Finite Mathematics*, ninth edition, and *Finite Mathematics and Calculus with Applications,* eighth edition. Chapter 1 is common to both texts, as well as *Calculus with Applications*, ninth edition, and *Calculus with Applications: Brief Version*, ninth edition. Part II is organized by chapters in *Finite Mathematics*; since not all chapters require detailed explanations of graphing calculator use, some chapters are not mentioned here. Instructions are given first for the TI-83 and 83/84 Plus, followed by instructions for the TI-85, TI-86, and TI-89, where appropriate.

In this manual, section titles from the textbooks are indicated in italics. References are made to specific examples and exercises from the corresponding sections of each chapter, so you should have your textbook nearby as you read through these instructions.

Chapter 1 Linear Functions

LOCATION IN THE OTHER TEXTS:

Calculus with Applications: Chapter 1
Calculus with Applications, Brief Version: Chapter 1
Finite Mathematics and Calculus with Applications: Chapter 1

Slopes and Equations for Lines.

Entering Lists and Plotting Points.

A set of points can be plotted, as in Example 15, by first entering the coordinates of the points into two lists--one list for the x-values and one for the y-values. On the TI-83 and 83/84 Plus, there are six list names to choose from--L_1 through L_6. The TI-83 and 83/84 Plus allow more lists to be added with user-defined names, as do the TI-85, TI-86 and TI-89.

The steps for entering and graphing lists of points is almost identical on the TI-83 and 83/84 Plus. Start by pressing [STAT] [ENTER] to obtain the list editor. Decide which list name will contain the x-values and use the left/right arrow keys to move the cursor to that list. If there is old data in the list that is no longer needed, use the up arrow to move the cursor to the list name and press [CLEAR] [ENTER] to delete the data. Enter the x-values, one at a time, in order, pressing [ENTER] after each. Choose a second list for the y-values and repeat the same process, being careful that corresponding x- and y-values are in the same position within each list (See Figure 1.)

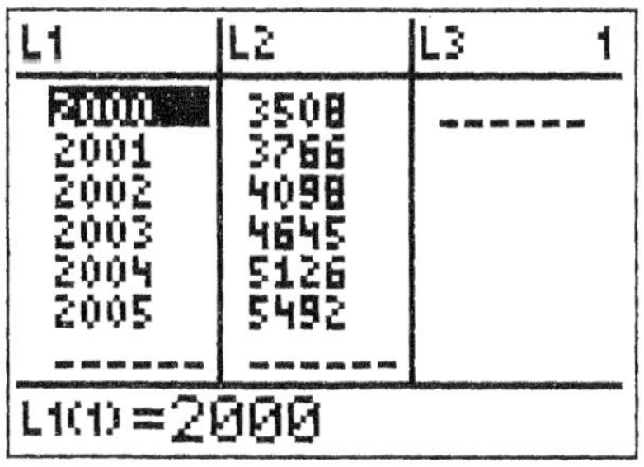

L1	L2	L3 1
2000	3508	------
2001	3766	
2002	4098	
2003	4645	
2004	5126	
2005	5492	
------	------	

L1(1)=2000

Figure 1.

To plot the data values, we must go through a few extra steps to make sure that the graph does not contain information we do not want. Begin by pressing [Y=] and "turning off" all functions stored there. To do this, move the cursor to the "=" beside any stored function and press [ENTER] to unhighlight the "=." Now, press [WINDOW] and enter the appropriate range values. To define the plot, press [2nd] [Y=] to obtain the **STAT PLOTS** menu. Press [4] [ENTER] to turn off any other plots, then return to the STAT PLOTS menu. Press [1], [2], or [3] to select a plot name. Turn the plot "on" by pressing [ENTER]. The cursor moves to the TYPE line; select the first plot type, the *scatter plot*, by again pressing [ENTER]. Next, the lines titled Xlist and Ylist need to contain the appropriate list names. On the TI-83 and 83/84 Plus, simply type in the name of the list containing the appropriate values. Finally, select one of three Marks that will represent the points on the graph. The plot is now defined; to see it, press [GRAPH]. (See Figure 2, which shows the data plotted in a [2000, 2006] by [3500, 5700] window.)

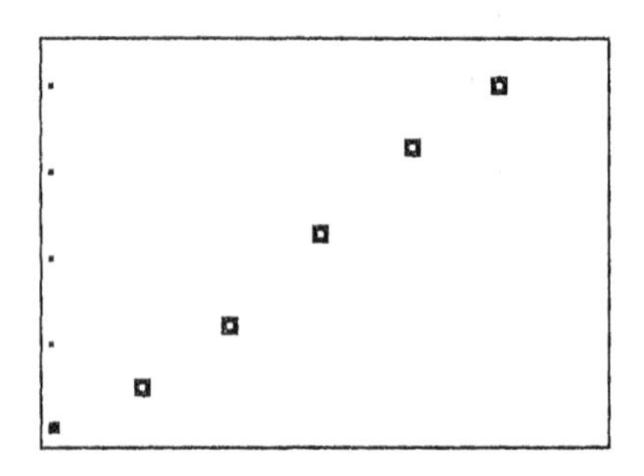

Figure 2.

Saving and graphing lists on the TI-85 is somewhat different. Begin by pressing [STAT] [F2] to obtain the list editor. Choose a name for the xlist by typing in a descriptive name and pressing [ENTER]; repeat for the ylist. For Example 15, appropriate list names might be "YEAR" and "COST". Once the list names have been entered, the calculator moves to the editing screen. Press [F5] to clear any old data values. Entering the points, one at a time, as follows: x-value, [ENTER], y-value [ENTER]. Before graphing the points, "turn off" any stored functions by pressing [GRAPH] [F1] [MORE] [F2]. Now press [GRAPH] [F2] and enter the appropriate range values. Finally, press [STAT] [F3] [F2] to draw the points.

If you are using the TI-86, press [2nd] [+] to access the STAT menu, and [F2] to EDIT. (Lists can be cleared in the same way as described for the TI-82 and TI-83 and 83/84 Plus.) Enter the x-values in the list xStat, one at a time, pressing [ENTER] after each. Use the right arrow key to move to yStat and enter the corresponding y-values, in order. To define a statistics plot, press [2nd] [+] to return to the STAT menu then [F3] to access the PLOT submenu. To define the plot as Plot1, press [F1] and press [ENTER] to turn the plot "on." Use the down arrow key to move to the Type line; press [F1] to select SCAT for the scatter plot. Press [ENTER]to move to the Xlist Name line; select xStat for the Xlist Name and press [ENTER] again. Select yStat for the Ylist Name and press [ENTER]. Choose one of three Marks to represent the points in the graphics screen. Turn off any stored functions by pressing [GRAPH] [F1] [MORE] [F2]. Now press [GRAPH] [F2] and enter the appropriate range values. Finally, press [F5] to see the graph.

On the TI-89, press [APPS] and select item 6: Data/Matrix Editor, then item 3: New. Next to the Type option, select Data. Type X for the Variable. Press [ENTER] twice. When the table is displayed, enter the x-values in the first column, pressing [ENTER] after each. Use the right arrow key to move to the second columns and enter the corresponding y-values.

When finished, press [F2] to enter the `Plot Setup` menu. To define how the plot will be displayed, press [F1], and choose `Scatter` as the `Plot Type`. Press the down arrow key and select one of five marks to represent the points in the graphics screen. Next to x, type `C1`, which is the name of the column containing the x-values. Next to y, type `C2`. Press [ENTER] twice to accept your changes. Make sure all unnecessary functions are turned off, and enter appropriate window values. Access the `GRAPH` command to see the scatter plot.

Linear Functions and Applications.

Finding Intersection Points Graphically.

In Example 2, of Section 1.2, Figure 11b shows a calculator-generated graph of a pair of supply and demand curves, along with the *equilibrium point*, or *intersection point* of the two graphs. Once the equations of the two functions have been entered into the calculator, all other functions have been "turned off," the appropriate range variables have been chosen, and the graph has been displayed, the calculator can find the intersection point.

On the TI-83 and 83/84 Plus, while the graph is displayed, press [2nd] [TRACE] to enter the `CALC` menu, then [5] for the `intersect` command. Press ENTER twice to select the two curves displayed. Use the left/right arrow keys to move the cursor as close to the intersection point as possible and press [ENTER] again to display the intersection point.

On the TI-85, while the graph is displayed, press [GRAPH] [MORE] [F1] to obtain the `MATH` menu, then [MORE] [F5] for the `isect` command. Press [ENTER] twice to select the two curves displayed; the intersection point will be displayed.

On the TI-86, while the graph is displayed, press [GRAPH] [MORE] [F1] to obtain the `MATH` menu, then [MORE] [F3] for the `isect` command. Press [ENTER] twice to select the two curves displayed; use the left/right arrow keys to move the cursor as close to the intersection point as possible and press [ENTER] again. The approximate coordinates of the intersection point will be displayed.

If you are using the TI-89, while the graph is displayed, press [F5] and choose the fifth option, `Intersection`. Press [ENTER] twice to select the two curves displayed. Use the left/right arrow keys to move to a point on the left of the intersection point and press [ENTER]. Move the cursor to a point on the right of the intersection point and press [ENTER] again. The approximate coordinates of the intersection point will be displayed.

Using Solving Equations.

After Example 2 of the text, the authors mention that graphing calculators can be used to solve equations. If an equation contains only one variable, then the `Solver` command on the TI-82, TI-85, and TI-86, or the `solve` function on the TI-89, can solve it.

Before executing the `Solver` command on the TI-83 and 83/84 Plus, the equation *must* be rewritten so that the left-hand side is "0." Once this bit of algebra is completed, press [MATH]

[0] to access the `solver` from the calculator homescreen. Press the up arrow, then type in the right-hand side of the equation. Type in a guess for the solution (see previous paragraph), and press [ALPHA] [ENTER] for the exact solution. (See Figure 3.) Consult the calculator guidebook for an explanation of the other items displayed.

```
9-1.5Q=0
▪Q=6
 bound={-1E99,1…
▪left-rt=0
```

Figure 3.

If you are using the TI-85 or the TI-86, press [2nd] [GRAPH] to obtain the `solver` menu. Press [CLEAR], then type in the equation. (Note: The TI-85 does *not* require that one side of the equation be "0.") Press [ENTER] and type in a guess for the variable. (See above for information about guesses.) Press [F5] for the solution.

On the TI-89, from the homescreen, press [2nd] [5] to access the `MATH` menu, and select option `9: Algebra`. Press [ENTER] to access the `solve` command, which is the first item in the `Algebra` menu. Type in the equation to be solved, followed by a comma, then type in the variable for which to solve and close the parentheses. Press [ENTER] to see the solution.

The Least Squares Line

Finding the Least Squares Regression Line.

To complete the example in this section, begin by entering the lists of points as described above. See the instructions beginning just before Figure 17 in Chapter 1 of your text for calculating the regression line on the TI-83 and 83/84 Plus.

If you are using the TI-85, follow the steps outlined previously for storing and displaying lists of points. To calculate the regression line, press [EXIT] [F1] to access the [STAT] **CALC** submenu. Type in the names of the lists containing the data, pressing [ENTER] after each. Press F2 to access and execute the `LINR` command; the regression line, $y = a + bx$, is calculated and the coefficients displayed.

If you are using the TI-86, after entering the lists, exit to the homescreen and press [2nd] [+] [F1] to access the **STAT CALC** menu, then [F3] for the `LinR` command. Following the command, type in the name of the x-list, a comma, then the name of the y-list. Press [ENTER] to see the coefficients for the least squares line.

If you are using the TI-89, after entering the lists, press [F5] to access the `Calc` menu. For the calculation type, choose option `5: LinReg`. Press the down arrow and type in the name of the column containing the x-values, then press the down arrow again and type in the name of the column containing the y-values. Press [ENTER] twice to see the coefficients for the least squares line.

Displaying Data and The Regression Line on the Same Graph.

Refer to the discussion above for entering and graphing lists of points. *After* the statistics plot has been defined and the calculator has found the least-squares regression line, the equation can be easily stored for graphing. On the TI-83 and 83/84 Plus, press [Y =] and move the cursor to down to an empty function name, keeping the cursor to the right of the "=" sign. Press [VARS] [5] for the `Statistics` variables menu. Press the right arrow twice to move to the `EQ` menu, and select `RegEQ`. The regression equation will be stored in the function menu and its function name will be turned "on." Alternatively, the command "`LinReg L1,L2,Y1`" automatically stores the regression equation under the function name `Y1`. (Other list names and function names may be used here, as necessary.) Press [GRAPH] to see the data and the regression line.

If you are using the TI-85, follow the steps outlined previously for storing and displaying lists of points. Next, calculate the regression equation as described above. Press [EXIT] twice to return to the homescreen, then turn "off" any unnecessary functions and set up appropriate range variables for the data in the [GRAPH] menu. *After* displaying the points, press [STAT] [F5] to draw the regression line.

If you are using the TI-86, after entering the lists, exit to the homescreen and press [2nd] [+] [F1] to access the `STAT CALC` menu, then [F3] for the `LinR` command. Following the command, type in the name of the x-list, a comma, the name of the y-list, another comma, then the function name under which to store the regression equation. For instance, "`LinReg xStat,yStat,y1`" calculates the regression line and stores its equation as "`y1`". Set up a statistics plot as described previously, define the range variables, and press [GRAPH] [F5] to see the scatter plot and the regression line.

If you are using the TI-89, after entering the lists, press [F5] to access the Calc menu. Set up the window as previously described, but choose a function name in which to store the regression equation, `RegEQ`. After choosing appropriate range values, access the `GRAPH` command to see the regression line and the scatter plot.

Chapter 2 Systems of Linear Equations and Matrices

LOCATION IN THE OTHER TEXTS:

Finite Mathematics and Calculus with Applications: Chapter 2

Solution of Linear Systems by the Gauss-Jordan Method.

Row Operations.

Once a matrix has been entered, the row operations required for the Gauss-Jordan method can be performed on the calculator. The steps for entering a matrix and performing the row operations are given below.

To enter the matrix of Example 2 on the TI-83, press [MATRX] and press the left arrow key to move to the `EDIT` menu. (On the TI-83/84 Plus, press 2nd MATRX.) There are five different matrix names to choose from on the TI-82 and ten on the TI-83 and 83/84 Plus. To begin editing matrix [A], press [1]. The dimension of the matrix must be entered first; type in the number of *rows*, press [ENTER], then the number of *columns*, and press [ENTER] again. In Example 2, there are 3 rows and 4 columns. Type in the matrix entries, from left to right, top to bottom, pressing [ENTER] after each one. Press [2nd] [MODE] to return to the homescreen.

To enter the same matrix on the TI-85 or TI-86, press [2nd] [7] to access the `MATRX` menu, then [F2] to begin the editing process. Type in a name for the matrix; for example, "`A`," and press [ENTER]. Follow the same steps as outlined in the previous paragraph for typing in the dimension and the individual matrix entries. Press [EXIT] to return to the homescreen.

On the TI-89, enter the matrix via the `Data/Matrix Editor` from the `APPS` menu. Choose `Matrix` for the Type. Type in a `Variable` name that is not currently in use; this will serve as the name of the matrix being stored. Press [ENTER] twice. Use the down arrow keys to move to the `Row Dimension` and type in the number of rows. Press the down arrow again and type in the `Column Dimension`. Press [ENTER]. Type in the elements of the matrix. Use the arrow keys to move between columns. Return to the home screen.

The row operations for completing the Gauss-Jordan process are located in the [MATRX] `MATH` menu of the TI-83 and 83/84 Plus ([MATRX] followed by the right arrow key), and in the `MATRX OPS` ([2nd] [7] [F4] [MORE]) menu of the TI-85 and TI-86. They are located in the `MATH`,

`Matrix, Row Ops` menu of the TI-89. The individual command names, their locations within the above menus, and their results are summarized below.

Interchange two rows

TI-83: rowSwap	Location:	Option C
TI-85: rSwap	Location:	F2
TI-86: rSwap	Location:	F2
TI-89: rowSwap	Location:	Option 1

Multiply a row by a nonzero number

TI-83: *row	Location:	Option E
TI-85: multR	Location:	F4
TI-86: multR	Location:	F4
TI-89: mRow	Location:	Option 3

Add two rows

TI-83: row+	Location:	Option D
TI-85: rAdd	Location:	F3
TI-86: rAdd	Location:	F3
TI-89: rowAdd	Location:	Option 2

Add a nonzero multiple of one row to another row

TI-83: *row+	Location:	Option F
TI-85: mRAdd	Location:	F5
TI-86: mRAdd	Location:	F5
TI-89: mRowAdd	Location:	Option 4

Since the syntax for using commands grouped together above is the same for all TI models, we will work through Example 2 using the TI-83 steps.

Once the matrix has been entered, and the calculator has been returned to the homescreen, the matrix name can be copied to an expression or command. On the TI-83, first press MATRX, followed by the number corresponding to the name of the matrix. (On the TI-83/84 Plus, press 2nd MATRX. On the TI-85, TI-86 and TI-89, matrix names can be typed in directly from the keypad of the calculator.) To get zeros in rows 2 and 3 of the first column, use the following sequence of commands, pressing ENTER after each:

*row+(-3,[A],1,2)→[A]	Multiplies row 1 of [A] by -3; adds the result to row 2.
*row+(-1,[A],1,3)→[A]	Multiplies row 2 of [A] by -1; adds the result to row 3.
*row(1/6,[A],2))→[A]	Multiplies row 2 by 1/6.
row+([A],2,1)→[A]	Adds row 2 to row 1 of [A].
*row(-4,[A],2,3)→[A]	Multiplies row 2 of [A] by -4; adds the result to row 3.
*row(1/6,[A],2))→[A]	Multiplies row 2 by 1/6.

At this point, with all TI-models except the TI-89, the matrix contains lengthy decimals; care must be taken to avoid round-off errors. Now, or at any time that decimals appear in a calculation, we can convert the matrix entries into fractions by typing MATRX, pressing the number corresponding to the name of the matrix, then MATH 1 ENTER. (See Figure 1.)

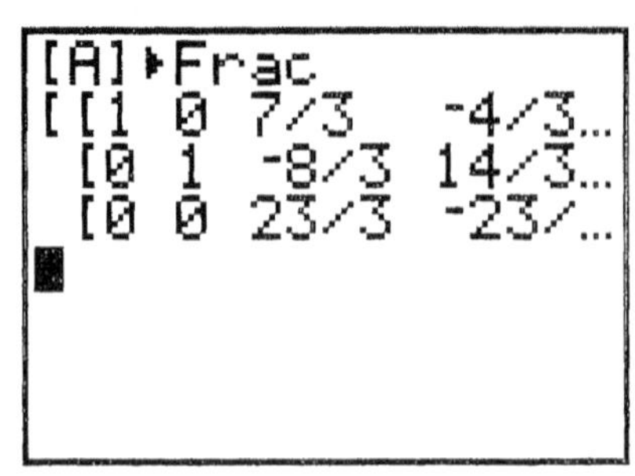

Figure 1.

To complete the Gauss-Jordan method, we use the following commands, pressing ENTER after each:

*row(3/23,[A],3)→[A]	Multiplies row 3 of [A] by 3/23.
[A]4Frac	Converts the matrix entries into fractions.
*row+(-7/3,[A],3,1)→[A]	Multiplies row 3 of [A] by -7/3; adds the result to row 1.
[A]4Frac	Converts the matrix entries into fractions.
*row+(8/3,[A],3,2)→[A]	Multiplies row 3 of [A] by 8/3; adds the result to row 2.

The `Frac` command is the first option in the MATH menu on the TI-83 and 83/84 Plus; it is in the MATH `MISC` menu of the TI-85 and TI-86. The final result is displayed in Figure 2.

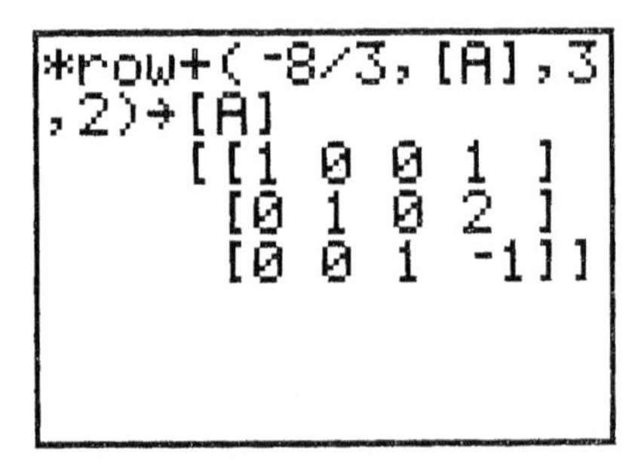

Figure 2.

The `rref` command.

As mentioned after Example 2 of the text, many calculators have a single command which will perform all of the steps of the Gauss-Jordan process at once. First, of course, the matrix must be entered into the calculator as described above.

Once the matrix has been entered into the TI-83 or 83/84 Plus, press MATRX, press the right arrow key to move to the `MATH` menu, and select option `B`; enter the matrix name from the MATRX menu and press ENTER to execute the command.

If you are using the TI-85 or TI-86, press [2nd] [7], then [F4] to access the `OPS` menu, and [F5] to access the `rref` command; type in the name of the matrix and press [ENTER] to complete the process.

On the TI-89, access the `rref` command in the `MATH MATRIX` menu; type in the name of the matrix, close the parentheses and press [ENTER] to execute the command.

Matrix Inverses.

Calculating the Inverse of a Matrix.

Once a square matrix has been stored, as described previously, the inverse of the matrix, if it exists, can be calculated quickly with any of the TI calculators. From the homescreen, access the name of the matrix on the TI-83 by pressing [MATRX], followed by the number corresponding to the name of the matrix. (On the TI-83/84 Plus, press 2nd MATRX. On the TI-85, TI-86 and TI-89, matrix names can be typed in directly from the keypad of the calculator.) For all models except the TI-89, press [x^{-1}] [ENTER]. The `Frac` command can again be used to convert the matrix entries to fractions, when possible.

To find the inverse of a matrix on the TI-89, press [^], and type "`(-1)`". Press [ENTER] to calculate the inverse.

A Note About Round-off Errors and Matrix Inverses.

Occasionally, when performing calculations involving matrix inverses, one or more entries in the resulting matrix may look like " 1 E -10," or something similar. Keep in mind that this represents 1 times 10 to the -10th power, or 0.0000000001. A result like this in a problem that originally contained no numbers of this form is due to round-off error in the calculator. Usually, it is safe to treat any similar results as "0."

Input-Output Models.

Creating an Identity Matrix.

As mentioned after Example 2 in the text, graphing calculators can be used to determine production for an input-output model. Once matrices A and D have been entered into the calculator, the appropriate identity matrix can be created, and the expression $(I - A)^{-1} D$ can be calculated.

On the TI-83 and 83/84 Plus, an identity matrix can be created using the `identity` command, located in the [MATRX] `MATH` menu. Access this command by pressing [5]. The command must be followed by a number, *n*, which represents the number of rows of the desired identity matrix. Remember to follow this number with [)]. In Example 2, the appropriate identity matrix will have dimension 3×3, so we would type "`identity (3)`" for this matrix.

If you are using the TI-85 or TI-86, an identity matrix can be created by pressing [2nd] [7] to access the **`MATRX`** menu, then [F4] for the `OPS` menu and [F3] for the `ident` command. Type in the number of desired rows for the matrix to complete the command.

On the TI-89, an identity matrix can be created by access the `identity` command, which the sixth option in the `MATH Matrix` menu. When accessed, you need only enter the number of rows for the matrix, close the parentheses, and continue with the expression to be evaluated.

Chapter 5 Mathematics of Finance

LOCATION IN THE OTHER TEXTS:

Finite Mathematics and Calculus with Applications: Chapter 5

Simple and Compound Interest.

Compounded Interest on the TI-83 and 83/84 Plus.

The TI-83 is equipped with a `TVM Solver`, available by pressing [2nd] [x^{-1}] [ENTER]. On the TI-83/84 Plus, press APPS ENTER. TVM stands for *time-value-of-money*. This solver will be handy throughout Chapter 5 of your text. There are also several `FINANCE` commands available, many of which are discussed below. It is very important to remember that these `FINANCE` commands cannot be used until values have been entered for the variables in the `TVM Solver`. The variables are as follows:

- `N` Number of payment periods.
- `I%` The annual percentage rate, given as a percent.
- `PV` The present value of the account. If money is being paid *into* the account, `PV` is entered as a negative number; otherwise, `PV` is entered as a positive number.
- `PMT` The amount of each payment; if money is being paid *out*, `PMT` is entered as a negative number; if money is being *earned* or *received*, then `PMT` is entered as a positive number.
- `FV` Future value of the account.
- `P/Y` Number of payments per year.
- `C/Y` Number of compounding per year.
- `END/BEGIN` If payments are paid at the end of the compounding period, choose `END`. If payments are made at the beginning of the compounding period, choose `BEGIN`.

Example 4 of this section of the text can solved by entering the values shown in Figure 1:

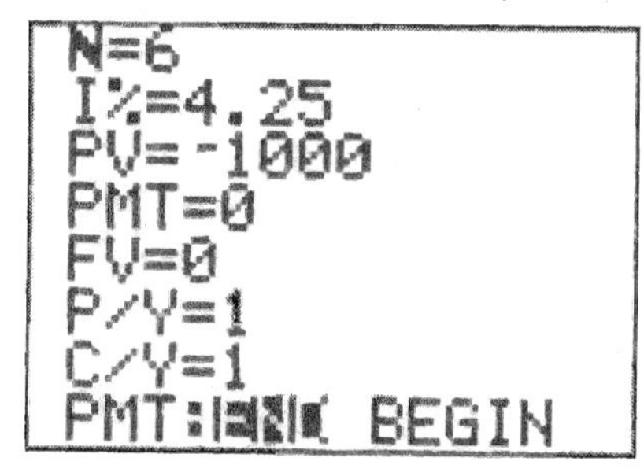

Figure 1.

In part (a), we are being asked to find the future value of the account, A. To do this with the solver, use the arrow keys to move the cursor beside FV and press [ALPHA] [ENTER] to execute the SOLVE command.

Note: *Often, the calculator's answers will differ slightly from those of the book.*

Effective Interest Rate on the TI-83 and 83/84 Plus.

Other commands in the FINANCE menu can be helpful in this section. The Eff command can be used to find the effective rate in Example 7. From the homescreen, press [2nd] [x^{-1}] and choose option C. (On the TI-83/84 Plus, press APPS and ENTER and choose option C. Type in the rate of compounded interest, *as a percent*, followed by a comma, then the number of compounding per year. Press [ENTER] to complete the command. (See Figure 2.)

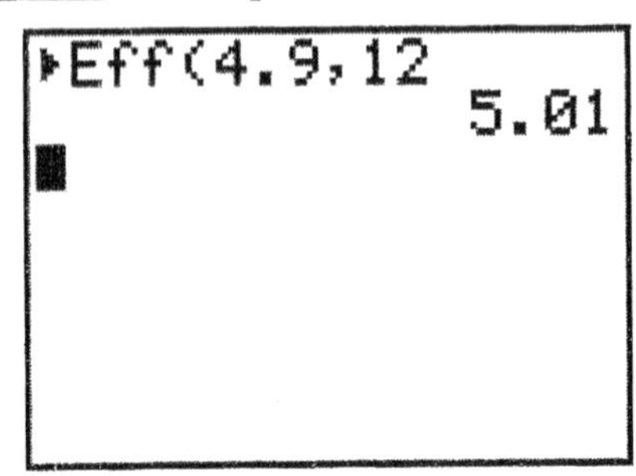

Figure 2.

Present value of an Account on the TI-83 and 83/84 Plus.

Example 9 can also be solved with the TVM Solver. Set **N** = 5, I% = 6.2, PV = 0 (since it is unknown), PMT = 0 (no additional payments will be made into the account), FV = 6000, P/Y = 1, C/Y = 1 (since interest is compounded annually), and choose END. Move the cursor beside PV and press [ALPHA] [ENTER] to find the amount to be deposited. (See Figure 3.)

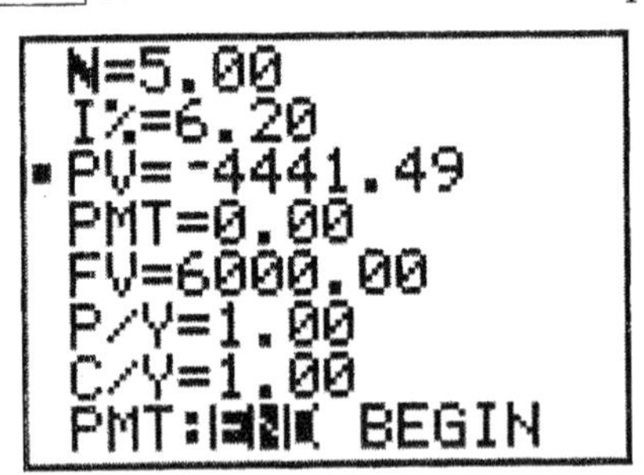

Figure 3.

Solving for N.

To solve Example 11 of Section 2, set I% = 8, PV = -1, PMT = 0, FV = 2, P/Y = 1, C/Y = 1, and choose END. Move the cursor beside **N** and press [ALPHA] [ENTER] to see that $1 will be worth $2 in approximately 9 years.

There are no built-in commands on the other TI models similar to those described in this section, however the functions can be programmed into the calculators. Alternatively, values can be found by entering the appropriate formula as an equation in the SOLVER menu of the TI-85 and TI-86, or the solve command on the TI-89. For instance, to complete Example 11 on the TI-85 or TI-86, press [2nd] [GRAPH] to obtain the SOLVER menu. Type in the equation to be solved, 2 = 1.08^N, and press [ENTER]. Type in a reasonable guess for the variable N and press [F5] to solve. To complete this example on the TI-89, access the solve command in the MATH Algebra menu, type in the equation, a comma, then the variable name, n. Close the parentheses and press ◆ [ENTER]. Also, a FINANCE app for the TI-89 can be downloaded from the TI website.

Future Value of an Annuity.

Ordinary Annuities.

To solve Example 4 of Section 2 we enter the values shown in Figure 4. Solving for FV, the future value of the given sinking fund can be determined.

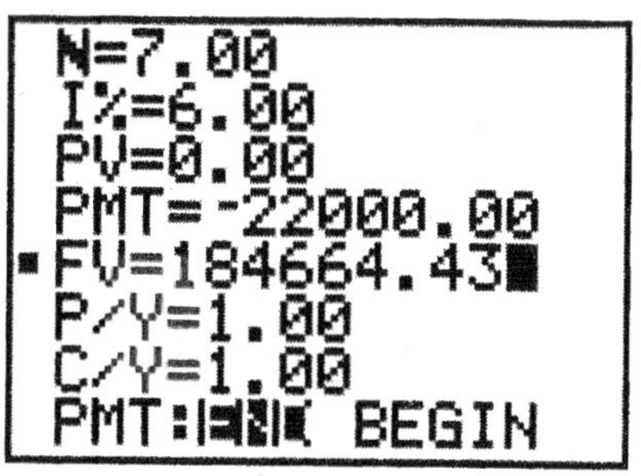

Figure 4.

To solve Example 5(a) with the TI-83 or 83/84 Plus, set N = 12×20 = 240, I% = 7.2, PV = 0 (since it is irrelevant here), PMT = -200, FV = 0 (since it is unknown), P/Y = 12, C/Y = 12 (since interest is compounded monthly), and choose END. Move the cursor beside FV and press [ALPHA] [ENTER] to complete the problem. For part (b) of the same example, replace FV by 130000, and solve for I%. Example 6 can be solved by returning I% to 7.2 and solving for PV.

The SOLVER menu of the TI-85 or TI-86, or the solve command on the TI-89, can also be used to solve parts (a) and (b) of Example 5, as previously described. Programs for the these models to calculate the future value of an ordinary annuity are included in Part III.

Annuities Due.

To solve this type of problem, select BEGIN in the last line of the TVM Solver. For instance, Figure 5 represents Example 7 of the text:

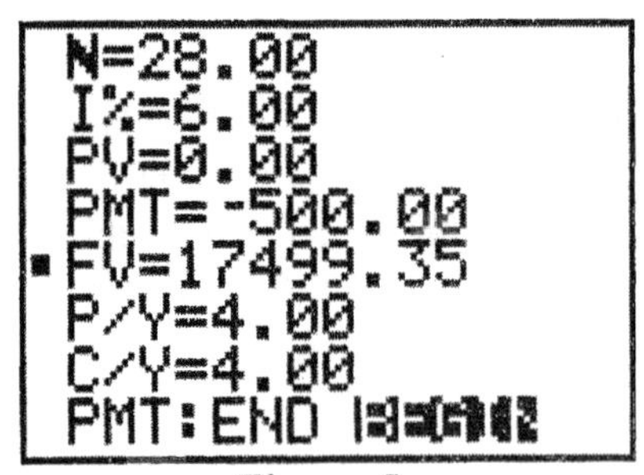

Figure 5.

Again, the other TI models do not have built-in functions to perform the financial calculations of this section. However, in Part III of this manual, programs for those models are available for calculating the future value of an annuity. The `SOLVER` menu of the TI-85 or TI-86, or the `solve` command on the TI-89, can also be used as described previously. This option is also on the FINANCE app for the TI-89 that can be downloaded from the TI website.

Present Value of an Annuity; Amortization.

Present Value of an Annuity on the TI-83 and 83/84 Plus.

The `TVM Solver` can again be used to perform the calculations necessary in this section of the text. To solve Example 1 of Section 3, press [2nd] [x^{-1}] [ENTER] to obtain the `TVM Solver`, and enter the values shown in Figure 6. Note that `FV` is set equal to `0` since it is irrelevant to the problem. Move the cursor beside `PV` and press [ALPHA] [ENTER] to complete the calculation.

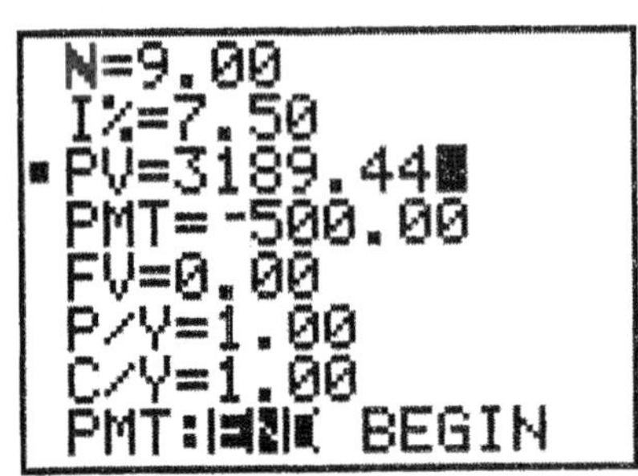

Figure 6.

Amortization.

The `TVM Solver` and other `FINANCE` functions of the TI-83 and 83/84 Plus can be used to calculate the size of the periodic payments required to amortize a loan, as well as to find the total amount of interest paid, and to create a partial amortization schedule. The other TI models can generate graphical representations of an amortization schedule, and the TI-86 or TI-89 can generate a partial amortization schedule; other functions can be programmed into the calculator. (See Part III.)

To complete Example 3 (a) with the TI-83 or 83/84 Plus, access the TVM Solver. Set N = 12×30 = 360, I% = 9.6, PV = 78000 (the size of the mortgage), PMT = 0 (since this is unknown), FV = 0 (since it is irrelevant to this problem), P/Y = 12, C/Y = 12 (since interest is compounded monthly), and choose END. Move the cursor beside PMT and press [ALPHA] [ENTER] to see that the payment size needs to be $661.56 per month in order to pay off the mortgage in 30 years. For part (b), exit to the homescreen by pressing [2nd] [MODE]. Press [2nd] [x^{-1}] to obtain the FINANCE menu, and select option A, Σint. This function computes the sum of the interest paid between any two payments. Since, in part (b), we want the total interest paid throughout the 30-year mortgage, we need to apply the function to the 1st through 360th payments. So, press [1], then a comma, then [3] [6] [0] and press [ENTER]. (See Figure 7.)

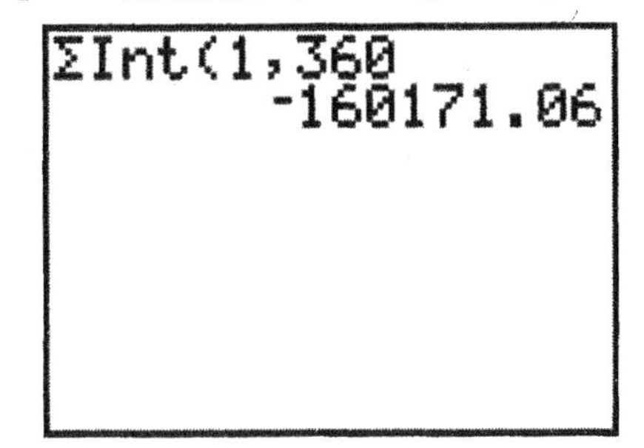

Figure 7.

Remember that the negative sign indicates that the interest is being *paid*, not *earned.* Part (c) can be solved with the same function, this time finding the amount of interest between the 1st and 1st payments; that is, evaluate Σint(1, 1). To find the remaining balance on the loan at any given payment, as described in the discussion following Example 3, we can use the bal(function. From the homescreen, press [2nd] [x^{-1}] to obtain the FINANCE menu, and select option 9. Enter the payment number, in this case, 60, and press [ENTER]. The unpaid balance after 5 years, or 60 payments, is $75,122.00.

The simplest way to generate the graph shown as Figure 12(b) of the text will also allow us to generate a partial amortization table for a given loan. To do this, we must change the function setting on the calculator. Press [MODE] and use the arrow keys to move the cursor to the word Par in the fourth line; press [ENTER]. The calculator is now set in *parametric mode.* (For more information about this mode setting, see the calculator's guidebook.) Once the proper values have been stored in the TVM Solver, press [Y=]. Notice that new function names appear there. Before proceeding, make sure that all Plots are "turned off." If a Plot is "on", it will be highlighted; turn it off by moving the cursor to the highlighted Plot name and pressing [ENTER]; repeat, if necessary, for other Plots. Move the cursor back to the first function name, "X_{1T}", and press [X,T,θ,n]; the letter "T" should appear. Move to the second function name, "Y_{1T}", and access the bal command by pressing [2nd] [x^{-1}] [9]. Press [X,T,θ,n] [)] to complete the definition. Here, X_{1T} = T will represent a payment number and Y_{1T} = bal(T) will represent the unpaid balance *after* that payment has been made.

Press WINDOW to set the range values for the graph as `Tmin` = 0, `Tmax` = 360, `Tstep` = 12, `Xmin` = 0, `Xmax` = 360, `Xscl` = 50, `Ymin` = 0, `Ymax` = 80000, and `Yscl` = 10000. (The `X` and `T` minimums and maximums should always be the same for this type of graph. The values were chosen to contain the first and last payments. `Tstep` was chosen as 12 so that each point on the graph will represent the balance at the end of 12 payments, or one year; this speeds up the graphing. The range for the `Y` values was chosen to include the size of the mortgage.) Press GRAPH. You may TRACE the graph to see the balance, `Y`, remaining at the end of each year during the 30-year period, `X`.

To see the remaining balance at the end of *each* payment, we can use the `TABLE` function of the calculator. Press 2nd WINDOW to access the `TBLSET` menu. Set `TblStart` = 0 and `ΔTbl` = 1. Make sure that `Auto` is highlighted for both `Indpnt` and `Depend`, and press 2nd GRAPH to see the partial amortization schedule. Scroll through the table with the up/down arrow keys to see the unpaid balance after each payment. (See Figure 8.) Figure 9 is obtained by pressing GRAPH.

T	X1T	Y1T
1.00	1.00	77962
2.00	2.00	77925
3.00	3.00	77886
4.00	4.00	77848
5.00	5.00	77809
6.00	6.00	77770
7.00	7.00	77731

T=1

Figure 8.

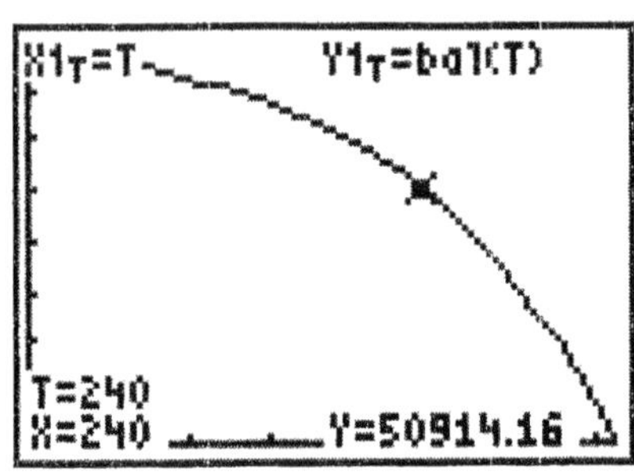

Figure 9.

Similar graphs can be obtained on the TI-85, TI-86 and TI-89, though the `TABLE` function is not available on the TI-85. Put the calculator into *parametric* mode as described above, and set the range variables as indicated. Define X1T (or `xt1` on the TI-89) as `T`. Because neither of these models contains the built-in `bal` command, define Y1T as indicated in the text, $R\left(\frac{1-(1+i)^{-(n-T)}}{i}\right)$; in Example 3, this would be Y1T $= 661.56\left(\frac{1-(1.008)^{-(360-T)}}{.008}\right)$. To view the partial table on the TI-86 or TI-89, access the `TBLST` or `TblSet` menu and store the values indicated previously. Access the `TABLE` command to view the table.

Chapter 8 Counting Principles; Further Probability Topics

LOCATION IN THE OTHER TEXTS:

Finite Mathematics and Calculus with Applications: Chapter 8

The Multiplication Principle; Permutations.

Factorial.

Factorial calculations, unless part of another set of commands or a program, should begin from the homescreen. Input the value of *n*, access the factorial, and press [ENTER].

For example, to calculate 10! on the TI-83 or 83/84 Plus, from the homescreen, type "`10`", then press [MATH], the left arrow key, then [4] [ENTER].

On the TI-85 or TI-86, the factorial is located in the `MATH PROB` menu. To calculate 10!, from the homescreen, type "`10`," then press [2nd] [×] to access the `MATH` menu, [F2] to access the `PROB` submenu, and [F1] for "`!`". Press [ENTER] to execute the command.

To calculate 10! with the TI-89, from the homescreen, type "`10`", then press [2nd] [5] to access the MATH menu, [7] to access the Probability submenu, then [1] for "`!`".Press [ENTER] to execute the command.

Permutations.

The permutation command on each of the calculators is located in the same area as the factorial command. To calculate a permutation, begin from the homescreen, unless the permutation is part of another function or program. Enter the value of *n*, access the permutation command, enter the value of *r* and press [ENTER].

For instance, to calculate P(8, 3) with a TI-82 or TI-83 or 83/84 Plus, as in Example 6 of the text, begin from the homescreen and type "`8`". Press [MATH], the left arrow key, then [2] to access the `nPr` command. Type "`3`" and press [ENTER] to complete the calculation.

Using the TI-85 or TI-86, calculate P(8, 3) by typing "`8`" on the homescreen. Press [2nd] [×] [F2] to access the `MATH PROB` menu, and [F2] for the `nPr` command. Type "`3`" and press [ENTER] to finish.

To calculate P(8, 3) on the TI-89, from the homescreen press [2nd] [5] to access the `MATH` menu, [7] to access the Probability submenu, then [2] for "`nPr(`". Type "`8`", a comma, "`3`", then [)]. Press [ENTER] to execute the command.

Combinations.

The combination command is also located in the same area as the two commands above. Execute the combination command in the same way that you execute the permutation command.

Example 1, $\binom{8}{3}$, is completed on the TI-83 or 83/84 Plus by typing [8], then [MATH], the left arrow key, then [3] to access the `nCr` command, and [3] [ENTER].

On the TI-85 or TI-86, type [8], then [2nd] [×] [F2] [F3] for the `nCr` command, and [3] [ENTER].

To calculate $\binom{8}{3}$ on the TI-89, from the homescreen press [2nd] [5] to access the `MATH` menu, [7] to access the Probability submenu, then [3] for "`nCr(`". Type "8", a comma, "3", then [)]. Press [ENTER] to execute the command.

Binomial Probability.

The `binompdf` and `binomcdf` Commands on the TI-83 or 83/84 Plus.

The TI-83 and 83/84 Plus have several probability functions built-in, including the binomial distribution. The appropriate commands are located in the `DISTR` menu of the calculator. The `binompdf` command is option `0`. The command requires you to input the number of trials, n, followed by the probability, p, of one "success," and the number of desired "successes," x. When executed, this command returns the probability of x successes in n trials. The `binomcdf` command requires the same input, but instead returns the probability of x or fewer successes in n trials.

To use these commands to complete Example 4, begin from the homescreen. In part (a), we are asked to find the probability of 1 defective item in a set of 15, where the probability of an item being defective is 0.01. Press [2nd] [VARS] to access the `DISTR` menu, then [0] to access the `binompdf` command. Type "`15`" followed by a comma, "`.01`" followed by another comma, then "`1`" for the number of desired "successes." Press [ENTER] to execute the command. Part (b) of Example 4 asks for the probability of *at most* 1 defective item in a set of 15. The `binomcdf` command will provide this information. It is option `A` in the `DISTR` menu. (See Figure 1.)

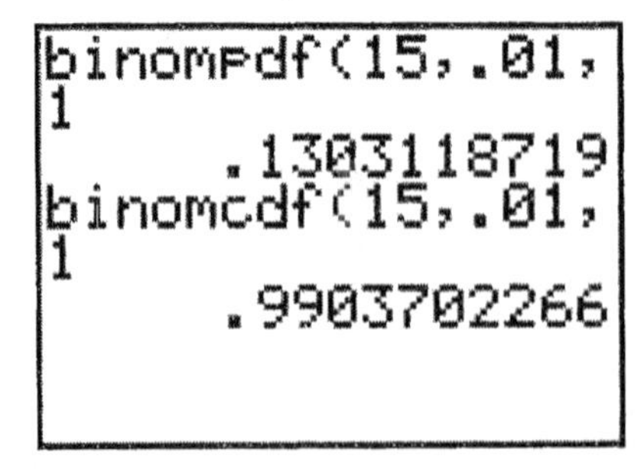

Figure 1.

Although the TI-85, TI-86 and TI-89 do not have a built-in binomial distribution function, it is not difficult to evaluate problems like Example 4(a) with these models, since the binomial probability function, $\binom{n}{x}p^x(1-p)^{n-x}$, can be typed directly onto the homescreen. On the TI-85 or TI-86, press (1 5, access the nCr command, then press 1) (. 0 1 ^ 1) (. 9 9 ^ 1 4) ENTER. The result will appear as in Figure 2.

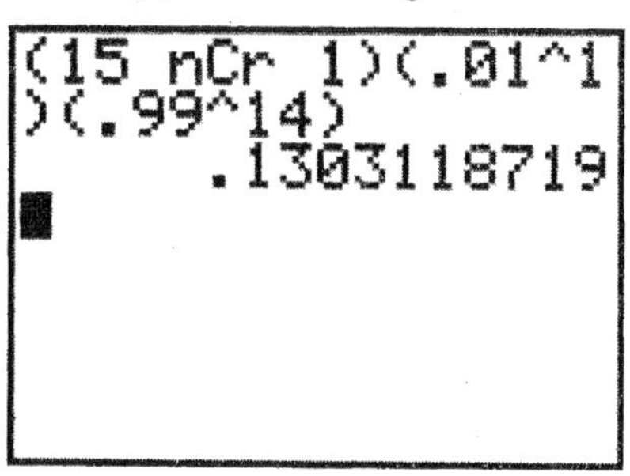

Figure 2.

With the TI-89, access the nCr command, then press 1 5, a comma, then 1) (• 0 1 ^ 1) (• 9 9 ^ 1 4) ENTER.

The TI-85, TI-86 and TI-89 can be programmed to complete problems such as Example 4(b). Also, there is a statistical package for the TI-89 that can be downloaded from the TI website that has a binomial distribution function built in.

Chapter 9 Statistics

LOCATION IN THE OTHER TEXTS:

Finite Mathematics and Calculus with Applications: Chapter 9

Frequency Distributions; Measures of Central Tendency.

Frequency Histograms.

The data in Example 1 of the text can be stored into and displayed by the graphing calculator. In this case, the data should be entered as a single list. (See discussion on page II-3 for assistance with storing lists.) The calculator will use the value input for `Xscl` as the width of the rectangles in a frequency histogram, so it is important to choose range variables that are appropriate for a given data set. In Example 1, `Xscl` is set equal to 5; thus, each rectangle in the histogram drawn by the calculator will be 5 units in length.

After storing the data list into the TI-83 or 83/84 Plus, we must define a statistics plot. Press [2nd] [Y=] and select a statistics plot name. Turn the plot "on;" beside `Type`, highlight the icon that resembles a histogram. Make sure that the list name under which the data is stored is chosen as the `Xlist`, and that "`1`" is chosen as the `Freq`. Turn "off" any functions saved in the [Y=] menu and enter the range values, indicated in the text, in the [WINDOW] menu. Press [GRAPH] to see the histogram. The graph can be traced to determine the height and range of each rectangle. (See Figure 1.)

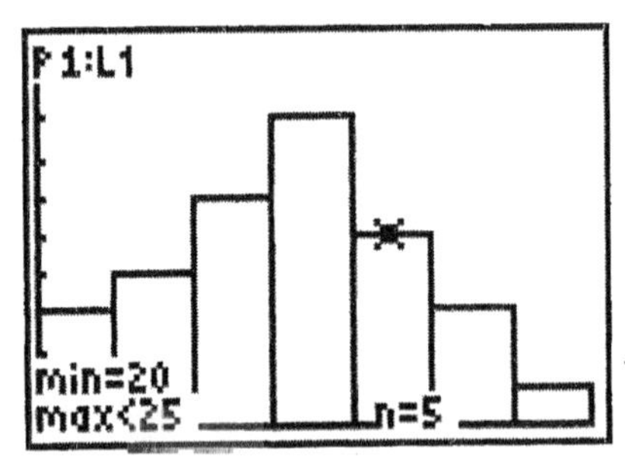

Figure 1.

If you have stored the data as a list in the TI-85, turn "off" any functions stored in the [GRAPH] `y(x)` menu and set the range variables as indicated in the text. Press [STAT] [F3] to access the `STAT DRAW` menu. If any old drawings appear on the graphics screen, press [F5] to clear them. Press [F1] to see the histogram. You cannot trace a histogram on the TI-85.

If you are using the TI-86, enter the data values into `xStat` and the corresponding frequencies into `fStat`. Set up the statistics plot and select `HIST` as the plot `Type` and `fStat` as the frequency `list`. Turn "off" any unnecessary functions, define appropriate range variables as described above, and press [GRAPH] [F5] to see the histogram. You may trace the histogram to view the height and range of each rectangle.

After storing the data and frequency in the `Data/Matrix Editor`, press [F2] to set up the plot. Choose a plot name, then press [F1] to define it. For the `Plot Type`, choose `Histogram`. Next to `x`, type in the name of the column containing the data values. Press the down arrow and type in a width for the rectangles in your histogram. Select `Yes` after `Freq and Categories`, to indicate that a frequency list has also been entered. Next to `Freq`, type in the name of the column containing the frequencies. Press [ENTER] to accept these changes. Turn off any unnecessary functions in the `Y=` menu, enter appropriate range values for the variables, then view the graph.

If you are using the TI-89, set up a data table as described on page II-4 of this manual. Enter the data values in column `C1` and, if necessary, the frequencies in column `C2`. Press [F2] to set up the plot, and choose `Histogram` as the `Plot type`. If you have a frequency column, make sure to enter it next to `Freq`. Display the histogram as described on page II-4.

For grouped data, a histogram can be created by storing the *midpoints* of the groups as one list (to be used as the **`Xlist`**), and the corresponding frequencies as another (to be used as **`Freq`**).

Frequency Polygons.

If data is entered as two lists, one containing the actual values, and the other containing the corresponding frequencies, then a frequency polygon can be drawn on the calculator. On the TI-83 or 83/84 Plus or TI-86, select the second plot **`Type`** and make sure that the appropriate lists are indicated for the data values and the frequencies. Select a **`Mark`** to represent the points. Finally, input the appropriate range values and view the graph.

On the TI-85, set up the appropriate range variables, press [STAT] [F3] to access the **`DRAW`** submenu, then [F3] to draw the **`xyline`**.

On the TI-89, from the `Data/Matrix Editor`, define the plot and choose `xyline` as the `Plot Type`, and `C1` for the x-values and `C2` for the y-values.. Set the `Freq and Categories` line to `No`. Otherwise, proceed as described previously.

Calculating Statistics.

Once a data set has been entered as a single list, it is an easy task to find its mean and median on the TI-83 or 83/84 Plus. From the homescreen, access the **`LIST`** menu by pressing [2nd] [STAT], use the left/right arrow keys to move to the **`MATH`** submenu, and select option 3, **`mean`**. Type in the name of the list which contains the data and press [ENTER]. The fourth option in the list, **`median`**, will find the median of a saved list. See Figure 2 on the next page for the mean and median of the data set from Example 1 of the text.

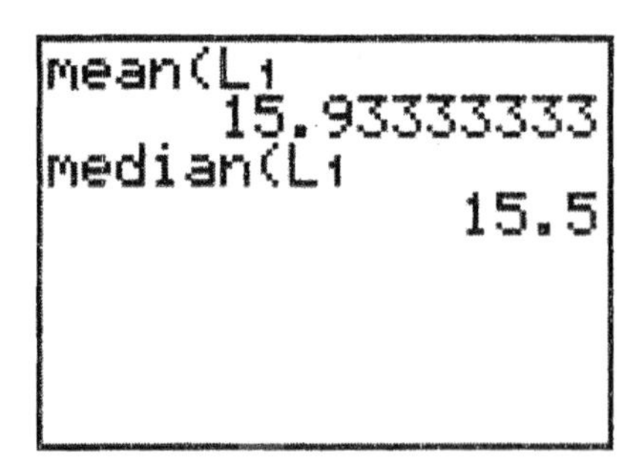

Figure 2.

If you are using the TI-85, the mean of a stored data set can be calculated by pressing [STAT] [F1] to access the `STAT CALC` menu; select the name of the list containing the data and press [ENTER]; then the name of the list containing the frequencies and press [ENTER] again. Press [F1] to access the `1-VAR` command, which calculates several one-variable statistics. The first statistic listed, $\bar{x}$, is the mean of the data set. The TI-85 does not have a command for calculating the median of a data set.

On the TI-86, press [2nd] [+] [F1] to access the `STAT CALC` menu, then [F1] for the `OneVar` command. Type in the name of the x-list, a comma, then the name of the y-list and press [ENTER]. The first statistic listed, $\bar{x}$, is the mean of the data set; use the down arrow key to scroll down to `Med`, which is the median of the data set.

On the TI-89, from the data screen, press [F5] and choose `OneVar` as the `Calculation Type`. Press the down arrow and enter the column name containing the data values. If there is a list containing frequencies, set `Freq and Categories` to `YES`, and enter the name of the list containing the frequencies next to `Freq`. Press [ENTER] to see the resulting calculations. The first statistic listed, $\bar{x}$, is the mean of the data set; `MedStat` is the median of the data set.

Measures of Variation.

Once a list of data is stored in the calculator, the standard deviation and variance can be calculated quickly. It is very important to remember that the calculator does not know whether the data entered represents a *sample* or a *population*. For that reason, all calculators covered by this manual, *except* the TI-89, will calculate both the population and sample standard deviations. These are distinguished on the TI calculators by `σx` and `Sx`, respectively.

On the TI-83 or 83/84 Plus, if data is saved in a single list, then option `7`, `stdDev`, and option `8`, `variance`, in the `LIST` menu can be used to calculate sample standard deviation and sample variance of that list.

If other statistics are required, or if data is entered as two lists (one containing the frequencies), then the `1-Var Stats` command should be used on the TI-83 or 83/84 Plus. From the homescreen, press [STAT] and the right arrow key to move to the `CALC` submenu; press [1]. Enter the name of the list containing the data. If a frequency list has been stored as well, then type a comma followed by the name of the frequency list. Press [ENTER]. (See Figure 5 of your text.)

Notice that the variance is not included in the list of calculated statistics. Recall that the variance is the square of the standard deviation, so it can be obtained from the information given. The values calculated and shown in Figures 3 and 4 are saved until you either perform a new statistical calculation or reset the calculator. To access these values, press [VARS] [5]. The third and fourth options listed there are the sample and population standard deviations, respectively. To calculate the sample variance from the sample standard deviation, press [3] [x^2] [ENTER].

The methods described previously for calculating the mean on the TI-85, TI-86 and TI-89 also calculate the sample standard deviations of a data set. The values calculated are stored by the TI-85 and TI-86 temporarily. After executing the `1-VAR` command on the TI-85, press [EXIT] twice to exit to the homescreen. Press [STAT] [F5] to access the `VARS` menu. To calculate the sample variance from the sample standard deviation, press [F3] [x^2] [ENTER].

After executing the `OneVar` command on the TI-86, press [EXIT] [F5] to access the `VARS` menu; calculate the sample variance as described for the TI-85.

The Normal Distribution.

The TI-83 and 83/84 Plus come equipped with commands to calculate normal probabilities, to find z-scores, and to graph the normal curve. While the TI-85, TI-86 and TI-89 do not have such built-in functions, they can be programmed to perform the calculations (see Part III); the equation for the normal probability function, $f(x)=\frac{1}{\sigma\sqrt{2\pi}}e^{-(x-\mu)^2/(2\sigma^2)}$, can be graphed by entering it directly into the [Y=] menus of the calculators, with appropriate values substituted for μ and σ.

To calculate the probability $P(a \le z \le b)$ on the TI-83 or 83/84 Plus, use the `normalcdf` command, which is option 2 in the `DISTR` menu. This command must be followed by the left-hand endpoint, a, a comma, then the right-hand endpoint, b. To calculate $P(-1.02 \le z \le 0.92)$, as in Example 1 (c) of the text, we would enter `normalcdf(-1.02,0.92)` on the homescreen and press [ENTER]. (See Figure 5.)

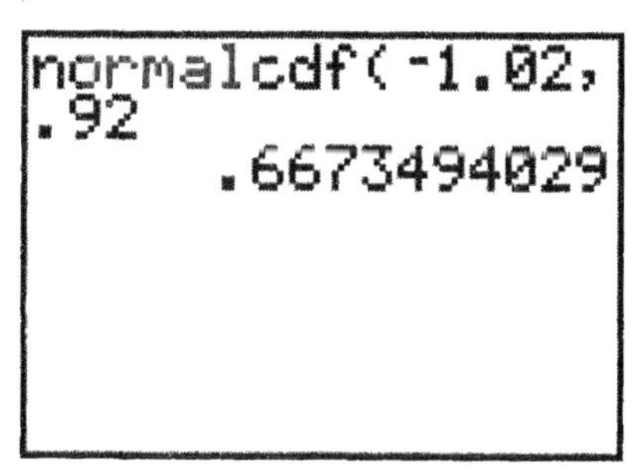

Figure 5.

To see a graph of the region, as well as to calculate the probability, we can use the `ShadeNorm` command in the `DISTR DRAW` menu. First, we should turn off all unnecessary functions and plots in the [Y=] and `STAT PLOT` menus. Next, we must set an appropriate range for the graph; for the *standard* normal distribution, we might set `Xmin` = -5, `Xmax` = 5, `Xscl` =

1, `Ymin` = -.1, `Ymax` = .45, and `Yscl` = .1. You may also want to press [2nd] [PRGM], followed by [ENTER] twice to clear any old graphs or drawings. Now, press [2nd] [VARS] to access the `DISTR` menu and press the right arrow key to move to the `DRAW` submenu. Press [1] to access the `ShadeNorm` command. As with the `normalcdf` command, we must follow this with the left-hand endpoint, a comma, and the right-hand endpoint. Press [ENTER] for the graph and the probability. (See Figure 6.)

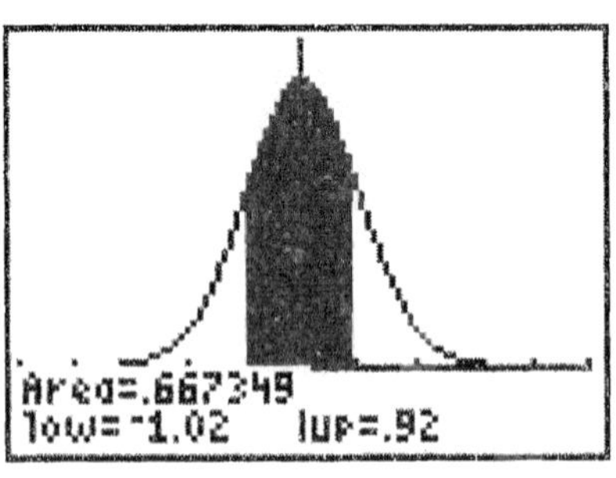

Figure 6.

For problems of the form $P(a \le z)$ or $P(z \le b)$, the calculator still requires both a left- and a right-hand endpoint in the `normalcdf` and `ShadeNorm` commands. In Example 1(a), we are asked to find $P(z \le 1.25)$; since no left-hand endpoint is given, we should choose a very large negative number (a number to the far left of the graph) to use in the calculator command. Your text suggests using -1×10^{99} in this case. Example 1(b) asks us to find $P(1.25 \le z)$; since no right-end point is given, we should choose a very large positive number to use in the calculator command. In such a case, your text suggests using 1×10^{99}. (See Figures 7 and 8.)

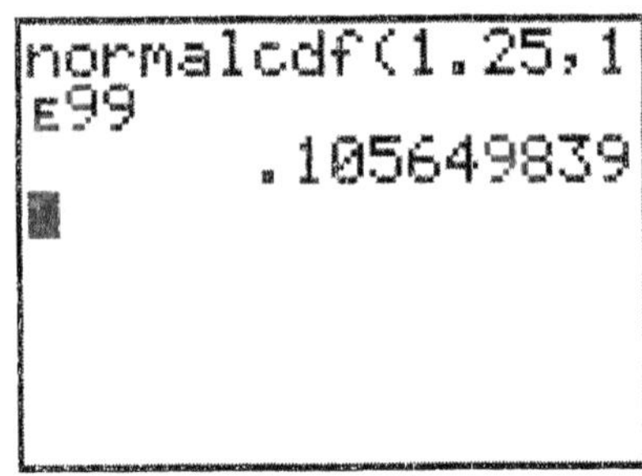

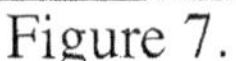
Figure 7.

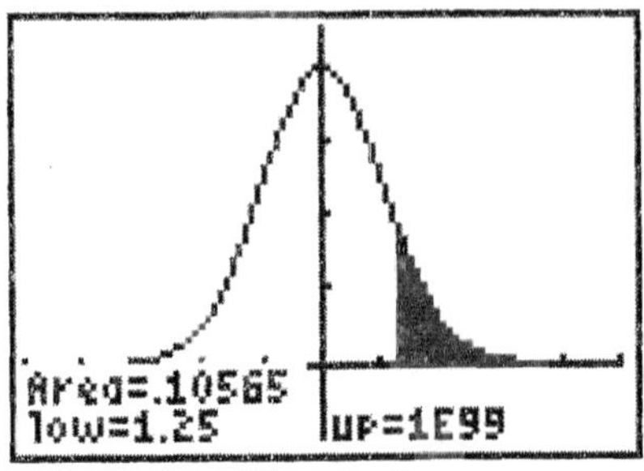

Figure 8.

All three commands discussed here can be used for any normal distribution by adding the mean and standard deviation of the distribution to the commands. For instance, Example 4(a) can be solved by using the command shown in Figure 9.

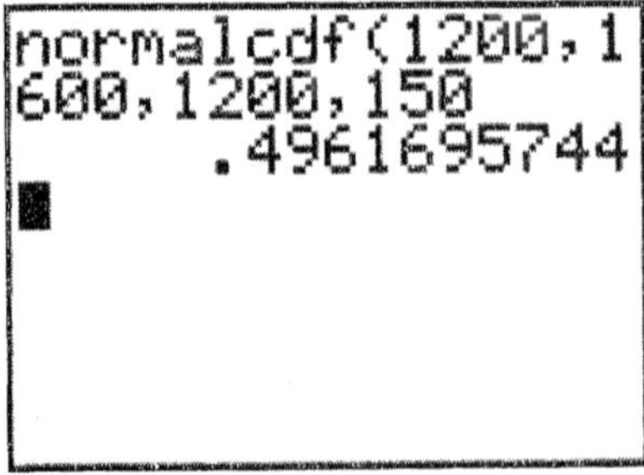

Figure 9.

In order to use the ShadeNorm command for other normal distributions, we must be careful to select an appropriate range. For Xmin and Xmax, choose values 4 or 5 standard deviations below and above the mean. To determine Ymax, we can use the command normalpdf to find the y-value of the normal distribution curve at its highest point (for a normal distribution, this location is *always* the mean). For the Example 4(b), press [2nd] [VARS] [1] to access this function, and type "1200,1200,150". These numbers represent the location of the highest point on the graph, the value of the mean, and the value of the standard deviation for the distribution; press [ENTER]. Use a value slightly above the resulting number as Ymax. Ymin can always be chosen as 0, if you do not mind having text cover a portion of the graph.

If you are using the TI-85, TI-86 or TI-89, probabilities involving the normal distribution can be approximated using the fnInt command. This command approximates the area under a curve which lies between two given x-values. From the homescreen, it can be accessed on the TI-82 by pressing [MATH] [9]. On the TI-85 and TI-86, the command is accessed by pressing [2nd] [÷] [F5]. On the TI-89, use the second option in the MATH Calculus menu. Follow this command with the appropriate formula for the normal distribution, $f(x)=\frac{1}{\sigma\sqrt{2\pi}}e^{-(x-\mu)^2/(2\sigma^2)}$, then a comma, the variable x, another comma, the left-hand x-value, a comma, and finally the right-hand x-value, close the parentheses and press [◆] [ENTER]. Example 1(b), $P(1.25 \le z)$, of the textbook can be approximated as shown in Figure 10. Note that the value 100 was used as the right-hand endpoint, and not 1E99. This is because using too large a number will result in an overflow error on the calculator when it attempts to square the number. Always use a value at least 5σ away from the mean when solving problems similar to Example 1(b) with these calculators.

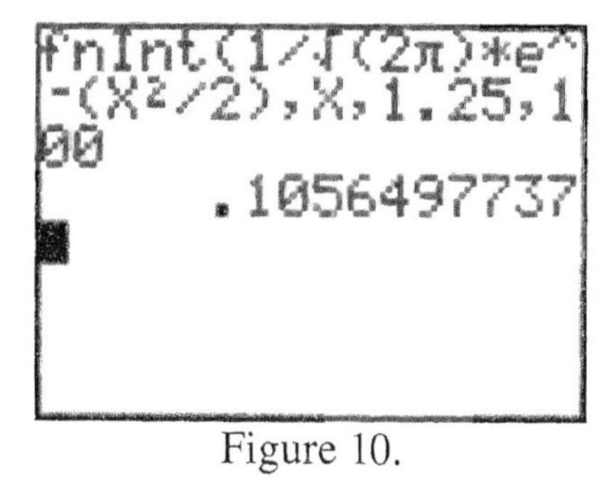

Figure 10.

Also, there is a statistical package for the TI-89 that can be downloaded from the TI website that has a normal distribution function built in.

Detailed Instructions for Calculus with Applications

This section contains detailed instructions for using the TI-83 or 83/84 Plus, TI-85, TI-86 and TI-89 with *Calculus with Applications,* ninth edition, *Calculus with Applications: Brief Version*, ninth edition, and *Finite Mathematics and Calculus with Applications*, eighth edition. (Instructions for Chapter 1 of all three texts begin on page II-3.) The section is organized by chapters in *Calculus with Applications*; since not all chapters require detailed explanations of graphing calculator use, some chapters are not mentioned here. Instructions are given first for the TI-83 or 83/84 Plus, followed by instructions for the TI-85, TI-86, and TI-89, where appropriate.

In this manual, section titles from the textbooks are indicated in italics. References are made to specific examples and exercises from the corresponding sections of each chapter, so you should have your textbook nearby as you read through these instructions.

Chapter 2 Nonlinear Functions

LOCATION IN THE OTHER TEXTS:

Calculus with Applications, Brief Version: Chapter 2
Finite Mathematics and Calculus with Applications: Chapter 10

Properties of Functions.

Evaluating Functions.

In Example 4(a) of the text, we are asked to calculate $g(3)$, where $g(x) = -x^2 + 4x - 5$. We can check our answers to this and similar problems using the calculator. First, enter the function as Y_1 and return to the homescreen by pressing [2nd] [MODE] to `QUIT` on the TI-82 and TI-83 and 83/84 Plus, or [EXIT] on the TI-85 or TI-86, or [HOME] on the TI-89.

To continue with the TI-83 or 83/84 Plus, press [VARS], followed by the right arrow key to move to the `Y-VARS` submenu. Press [ENTER] to access the `FUNCTION` submenu, then press the number corresponding to the function name under which $g(x)$ was stored. Type "`(3)`" and [ENTER] again to evaluate the function. (See Figure 1.)

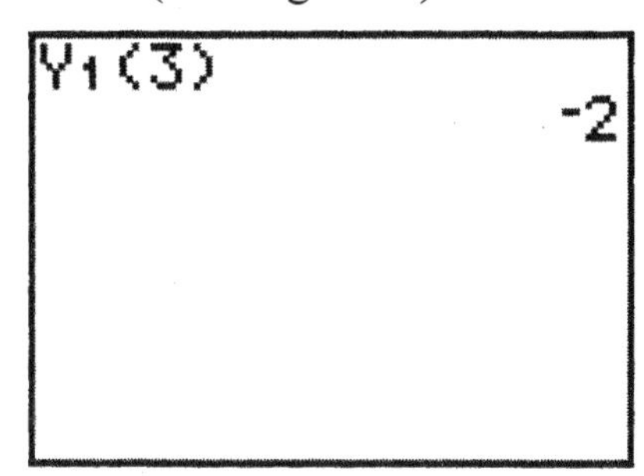

Figure 1.

Similarly, on the TI-82, once the function has been stored and the calculator has been returned to the homescreen, press [2nd] [VARS] to access the `Y-VARS` menu and proceed as described for the TI-83.

If you are using the TI-85 or TI-86, press [2nd] [÷] to access the `CALC` menu. Press [F1] for the `evalF` command. Type in the function name under which $g(x)$ was stored, followed by a comma, then the variable `x`, followed by a comma, and the x-value. Press [ENTER] to complete the calculation.

Using the TI-89, type "`y1(3)`" and press [ENTER] or ◆ [ENTER] to complete the calculation.

Evaluating Functions from the Graph Window.

While viewing the graph of a function on your calculator, you can evaluate the function at any x-value between `Xmin` and `Xmax`, inclusive. If you are using the TI-82 or TI-83 or 83/84 Plus, press [2nd] [TRACE] to access the `CALC` menu and choose the first command, `value`. The calculator returns to the graph and prompts for an x-value. To check our answer to Example 4(a) graphically, we would type "3" and press [ENTER]. The coordinates of the point on the graph where $x = 3$ are displayed and the point is marked.

If you are using the TI-85 or TI-86, press [GRAPH] [MORE] [MORE], then [F1] to access the `Eval` command. Type in the desired x-value and press [ENTER].

On the TI-89, while viewing the graph, press [F5] and choose `Value`. Type in the desired x-value and press [ENTER]. The coordinates of the point are displayed, and the cursor moves to that point on the graph.

Similarly, the intersection point of two functions can be found graphically. The instructions for doing this are located in the "Linear Functions" chapter of this manual, on page II-4.

Quadratic Regression.

In Exercise 59 of Section 2 of your text, a quadratic function is fit to a set of data representing Head Start enrollments. The quadratic regression feature of your graphing calculator (`QuadReg` on the TI-83 and 83/84 Plus and TI-89, `P2Reg` on the TI-85 and TI-86) is in the same location as the linear regression feature. First, data points given need to be entered into two lists—one for the x-values and one for the corresponding y-values. (See page II-3 of this manual for information on entering lists.) Once the data has been entered, return to the homescreen.

On the TI-82 and TI-83 and 83/84 Plus, press [STAT], and the right arrow key to move to the `CALC` submenu. Choose the `QuadReg` command. The command is copied onto the homescreen. Beside it, type the name of the list containing the x-values, a comma, then the name of the list containing the y-values and press [ENTER]. The coefficients for a quadratic function, $y = ax^2 + bx + c$, are calculated and displayed.

On the TI-85, press [STAT] [F1], then select the names of the lists containing the x-values and y-values. Press [MORE] and [F1] to choose the `P2REG` command, which calculates the coefficients for a quadratic function (second-degree polynomial) that best fits the data. Press [ENTER], and these coefficients are displayed in a list, $\{a, b, c\}$. You may have to scroll right, using the right arrow key, to view all of the coefficients.

On the TI-86, press [2nd] [+] [F1] to access the `STAT CALC` menu, then [MORE] [F4] to access the `P2Reg` command. Type in the name of the x-list, a comma, the name of the y-list and press [ENTER]. As with the TI-85, the coefficients of the best fitting quadratic function are displayed in a list, $\{a, b, c\}$. You may have to scroll right, using the right arrow key, to view all of the coefficients.

On the TI-89, follow the instructions for setting up a statistical calculation within the `Data/Matrix Editor`, as described for linear regression, but choose `QuadReg` as the `Calculation Type`. The values for a, b, and c, as well as for r^2, will be displayed.

If you are using the TI-83 or 83/84 Plus or the TI-86, you can automatically store the regression equation as a function for graphing by making an addition to the end of the command line. For instance, "`QuadReg L1, L2, Y1`" will calculate the quadratic regression equation and store it as `Y1` in the [Y=] menu. The regression equation can then be viewed with the scatter plot of the data. This same addition can be applied to any of the regression commands discussed in this manual.

Graphically Finding the Maximum or Minimum of a Function.

A graphing calculator can be used to determine the maximum of a function, such as the revenue function in part (b) of Example 7. Once the function has been stored, and the graph displayed, we can estimate the maximum (or minimum) for a particular function *within the chosen viewing window*. In the instructions given below, it is *very* important to select appropriate upper/lower bounds for the interval containing the desired maximum; if the interval between the chosen bounds is too wide, an incorrect answer may be obtained. In addition, it is also *very* important to move the cursor as close as possible to the desired maximum when selecting the "guess," when one is required by your calculator, otherwise the calculator may not give you the correct answer.

While viewing a graph on the TI-82 or TI-83 or 83/84 Plus, press [2nd] [TRACE] for the **`CALC`** menu. Select option `4, maximum`. You are asked to input a left bound for an interval which includes the maximum value of the function in the viewing window. Use the left arrow key to move the cursor to the left of the maximum and press [ENTER]. You are now asked for a right bound; use the right arrow key to move the cursor to the right of the maximum and press [ENTER]. Finally, you are asked to provide a "guess;" use the left/right arrow keys to move the cursor as close to the desired maximum as possible and press [ENTER]. An estimate for the x-value and the corresponding y-value are displayed.

If you are using the TI-85, press [GRAPH] [MORE], then [F1] to access the **`MATH`** submenu. Press [F1] again and move the cursor just to the left of the desired maximum and press [ENTER] to select a lower bound for an interval containing the maximum of the function; press [F2] and move the cursor just to the right of the desired intercept and press [ENTER] to select an upper bound. Press [MORE], then [F2] to obtain the **`FMAX`** command. Use the left/right arrow keys to move the cursor as close to the desired maximum as possible and press [ENTER] to see estimates for the maximum point's coordinates.

If you are using the TI-86, press [GRAPH] [MORE], then [F1] to access the **`MATH`** submenu. Press [F5] to obtain the **`FMAX`** command. Type in an x-value just slightly to the left of the desired maximum and press [ENTER]; type in an x-value slightly to the right of the desired maximum and press [ENTER] again. (Alternatively, use the left/right arrow keys to select these left/right endpoints.) Use the left/right arrow keys to move the cursor as close to the desired maximum as possible and press ENTER to see estimates for the maximum point's coordinates.

While viewing the graph of the function on the TI-89, press F5 and select Maximum. Move the cursor to the left of the point with the maximum y-value and press ENTER to select a lower bound, then move the cursor to the right of the point and press ENTER to select an upper bound. Press ENTER again to see the resulting coordinates of the point with the maximum y-value within the lower and upper bounds.

In order to estimate the minimum point of a function, within a chosen viewing window, follow the steps outlined above, but choosing minimum on the TI-82 and TI-83 and 83/84 Plus or FMIN on the TI-85 and TI-86.

Exponential Functions.

Exponential Regression.

Example 7(b) of this section of the text asks us to find an exponential function that models the given corn production. Another way to find an exponential function that fits a set of data is to use a graphing calculator or computer program with an exponential regression feature. Enter the years as one list in the calculator and the corresponding production levels as another list; return to the homescreen.

On the TI-82 and TI-83 and 83/84 Plus, press STAT, and the right arrow key to move to the CALC submenu. Choose the ExpReg command (option 0 on the TI-83 and 83/84 Plus and option A on the TI-82). The command is copied onto the homescreen. Beside it, type the name of the list containing the x-values, a comma, then the name of the list containing the y-values and press ENTER. The parameters for an exponential function, $y = a \cdot b^x$, are calculated and displayed. (See Figure 2.)

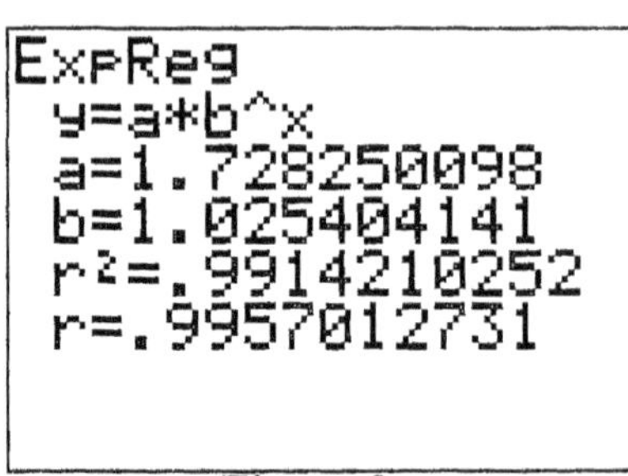

Figure 2.

On the TI-85, press STAT F1, then select the names of the lists containing the x-values and y-values. Press F4 to choose the EXPR command, which calculates the coefficients for a exponential function that best fits the data. Press ENTER, and these coefficients are displayed, along with the corresponding correlation coefficient.

If you are using the TI-86, press 2nd + F1 to access the STAT CALC menu, then F5 to access the ExpR command. Type in the name of the x-list, a comma, the name of the y-list and press ENTER.

If you are using the TI-89, set up the statistical calculation within the `Data/Matrix Editor`, as previously described, but choose `ExpReg` as the Calculation Type. Proceed as with other regression calculations.

Applications: Growth and Decay; Mathematics of Finance.

Power Regression.

In Exercise 105(d) from the review section of your textbook, you are asked to use the power regression feature of your calculator. This command will calculate the coefficient and exponent for the best-fitting function of the form $y = a \cdot x^b$. Follow the same steps as outlined above for the quadratic and exponential regression commands, this time choosing **PwrReg** on the TI-82 and TI-83 and 83/84 Plus, **PWRR** on the TI-85 or TI-86, or `PowerReg` on the TI-89.

Chapter 3 The Derivative

LOCATION IN THE OTHER TEXTS:

Calculus with Applications, Brief Version: Chapter 3
Finite Mathematics and Calculus with Applications: Chapter 11

Continuity.

Graphing Piecewise Functions.

As indicated in the discussion following Example 3, piecewise functions can be graphed on your graphing calculator. For all TI models *except* the TI-89 (see below), when storing the function under a function name, each piece of the function must be multiplied by a *test* function that indicates the interval corresponding to that piece, and all pieces should be added together. These test functions are located in the `TEST` menu of each calculator—[2nd] [MATH] on the TI-83 or 83/84 Plus; [2nd] [2] on the TI-85 and TI-86; and [2nd] [5] [8] on the TI-89. Use commands there to type in the piecewise function as it is written in your textbook.

To enter piecewise function on the TI-89, enter each piece as a different function, the use the | key to indicate where this function is used. For instance, y1 = (2x+3) | x< 2 indicates that this function is to be only for x<2.

Definition of the Derivative.

Graphing Tangent Lines.

The tangent line to a function at a given point may be graphed on the same screen as the graph of the function itself, as indicated in the note after Example 1 of your text. Store the function under a function name and turn off other functions and/or plots. Set the range variables so that the x-value of the point where the tangent line is to be drawn is *exactly* halfway between Xmin and Xmax. To duplicate Figure 32 of the text, you might choose `Xmin = -4` and `Xmax = 2`. Then graph the function with your calculator.

If you are using the TI-83 or 83/84 Plus, press [2nd] [PRGM] to access the `DRAW` menu and choose option 5, `Tangent`. The graphics screen is displayed again, with the blinking cursor at the center; because of the values chosen above for `Xmin` and `Xmax`, the cursor will already be located at the desired x-value. Otherwise, type the x-value of the point after accessing the `Tangent` command for the tangent line to the function at any point between `Xmin` and `Xmax`. Press [ENTER] for the graph of the tangent line. The TI-83 or 83/84 Plus will also display the x-value of the point and the equation for the tangent line at that point. (See Figure 1)

If you are using the TI-85, while viewing the graph of the function, press MORE F2 To access the DRAW menu, then MORE MORE F1 for the TanLn command. The calculator returns to the homescreen, where you should type the function name, followed by a comma, and the desired x-value of the location of the tangent line; press ENTER.

On the TI-86, while viewing the graph of the function, press MORE F2 to access the DRAW menu, then MORE MORE MORE F2 for the TanLn command. The calculator returns to the homescreen, where you should type the function name, followed by a comma, and the desired x-value of the location of the tangent line; press ENTER.

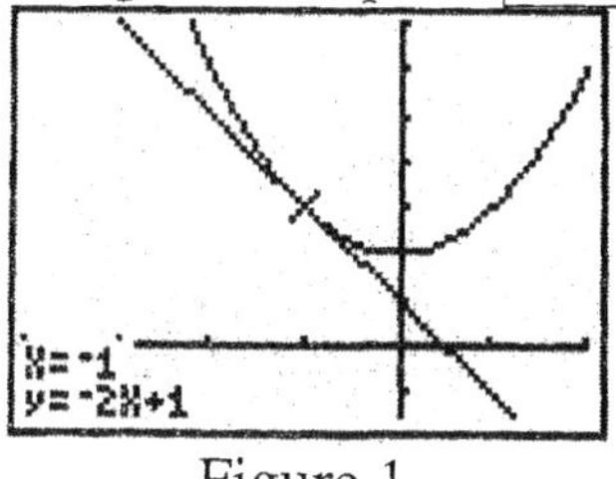

Figure 1.

If you are using the TI-89, while viewing the graph, press F5 and select option A: Tangent. Type in the x-value of the point whose tangent line you wish to draw.

Numerical Differentiation on the Calculator.

Graphing calculators are equipped with commands that calculate the derivative of a function, when it exists, for a given value of the variable. The TI-83 and 83/84 Plus, TI-85, and TI-86 all calculate derivatives numerically; hence these models only provide approximations of derivatives. The TI-89 calculates derivatives symbolically, much like you would by hand; hence this model provides exact values of derivatives.

If you are using the TI-83 or 83/84 Plus, from the homescreen press MATH and choose option 8, nDeriv. If you are using the TI-85 or TI-86, press 2nd ÷ to access the CALC menu, then F2 for the nDer command. Type in the function, followed by a comma, the variable, followed by another comma, and press ENTER. Figure 2 shows the result of Example 4(b) of this section of the textbook.

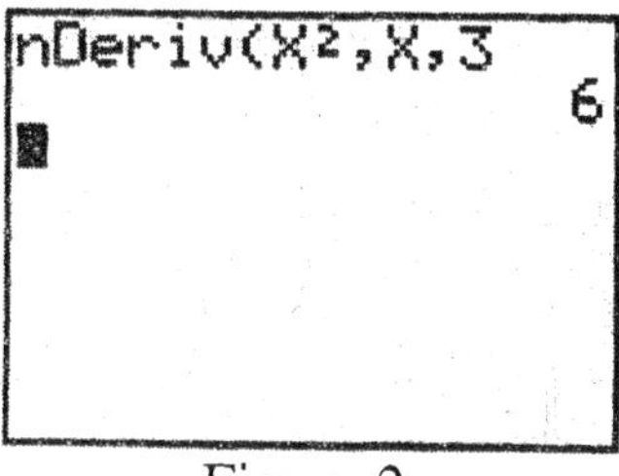

Figure 2.

The TI-85 and TI-86 also have a command which gives more exact values of derivatives for non-polynomial functions. This command, der1, is located in the same menu as the nDer command.

If you are using the TI-89, from the homescreen, press 2nd 8 to access *d*, the differentiation command. Type in the function (or function name), a comma, then the variable, and close the parentheses. Press ENTER to see the derivative of the function. If you wish to evaluate the derivative at a particular x-value, say $x = 3$ then immediately after the differentiation command, you can type |, then "x=3". Press ENTER for the exact value of the derivative.

While viewing the graph of a function, you can also find an approximation for the derivative of the function at a given point. First, store the function, and set the range values as described for graphing tangent lines. On the TI-83 or 83/84 Plus, press 2nd TRACE to access the CALC menu and choose option 6, dy/dx. On the TI-85 or TI-86, press MORE F1 for the MATH submenu, then F4 on the TI-85 or F2 on the TI-86 for the dy/dx command. On the TI-89, press F5, then select option 6: Derivatives. If you are using the TI-83 or 83/84 Plus, TI-86, or TI-89, you may also type the x-value of the point after accessing the dy/dx command, and press ENTER to calculate the derivative of the function at any point between Xmin and Xmax.

If you are using the TI-85, then move the cursor as close to the desired point as possible and press ENTER.

Chapter 6 Applications of the Derivative

LOCATION IN THE OTHER TEXTS:

Calculus with Applications, Brief Version: Chapter 6
Finite Mathematics and Calculus with Applications: Chapter 14

Absolute Extrema.

Checking Answers with the Calculator.

As discussed previously, the maximum and/or minimum of a function, within a certain range of x-values, can be approximated graphically with your calculator. Review the instructions on page II-33, if necessary, and note the importance of choosing appropriate upper and lower bounds for the intervals containing the desired extrema. Also, answers can be checked by using the evaluating the function on your calculator, as described on page II-32.

Applications of Extrema.

Approximating Critical Numbers Graphically.

Example 5 of this section involves the equation, $H'(S)=0$, or $2.17\left(\frac{\ln(S+1)}{2\sqrt{S}}+\frac{\sqrt{S}}{S+1}\right)-1=0$, which is quite difficult to solve. If $H'(S)$ is stored under a function name and graphed in a window containing the x-intercept(s), these points, called the *root(s)* or *zero(s)* of the function, can be approximated. If you are using the TI-83 or 83/84 Plus, while observing the graph, press [2nd] [TRACE] to access the `CALC` menu and choose option `2`, which is `zero` on the TI-83 or 83/84 Plus. Use the arrow keys to move the cursor to the left of the desired x-intercept and press [ENTER]. Use the right arrow key to move the cursor to the right of the desired intercept and press [ENTER] again. Finally, move the cursor as close to the intercept as possible and press [ENTER] once more; the approximate x-value of the intercept is displayed. (See Figure 2 on the next page.)

If you are using the TI-85, while observing the graph of the function, press [GRAPH] [MORE] [F1] to access the `MATH` submenu, then [F3] for the `root` command. Use the arrow keys to move the cursor as close to the desired intercept as possible and press [ENTER]; the approximate value will be displayed.

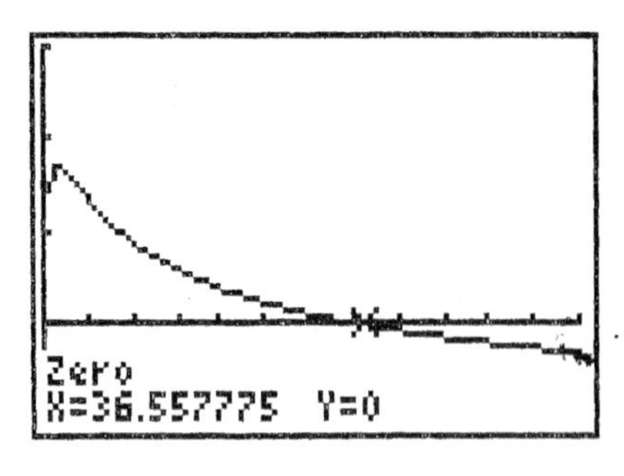

Figure 2.

On the TI-86, while observing the graph of the function, press [GRAPH] [MORE] [F1] to access the MATH submenu, then [F1] for the root command. Use the arrow keys to move the cursor to the left of the desired x-intercept and press [ENTER]. Use the right arrow key to move the cursor to the right of the desired intercept and press [ENTER] again. Use the arrow keys to move the cursor as close to the desired intercept as possible and press [ENTER] again; the approximate value will be displayed.

On the TI-89, while observing the graph of the function, press [F5] and select the second option, Zero. Use the arrow keys to move the cursor to the left of the desired x-intercept and press [ENTER]. Use the right arrow key to move the cursor to the right of the desired intercept and press [ENTER] again.

Alternatively, on the TI-83 and 83/84 Plus, TI-86 and TI-89, the left and right x-values surrounding the desired intercept can be typed in directly, without using the arrow keys, as can the "guess" on the TI-83 and TI-86.

Relative Extrema on the TI-89.

Since the TI-89 is capable of solving equations and differentiating symbolically, it can be used to find the critical numbers of a function. For example, if the function in question is stored as `y1`, then the following command will find the critical numbers of the function: "`solve(d(y1(x),x)=0,x)`".

Alternatively, to find the location of the relative maximum of a function, the command `fMax` in the `MATH Calculus` menu of commands can be used. If the function is stored as `y1`, then the command "`fMax(y1(x),x)`" will find the location of the relative maximum, if one exists. Similarly, the command `fMin` will calculate the x-value of the relative minimum of a function, if a minimum exists.

Chapter 7 Integration

LOCATION IN THE OTHER TEXTS:

Calculus with Applications, Brief Version: Chapter 7
Finite Mathematics and Calculus with Applications: Chapter 15

Area and Definite Integrals.

Summation.

The discussion following Example 1 demonstrates how lists can be used with the TI-83 and 83/84 Plus to gather the information necessary to approximate a definite integral and to calculate the approximation. To store the headings for list L2, after L1 has been stored, use the arrow keys to move the cursor to the top of the list, so that the list name, L2, is highlighted. Type the desired expression for the list (in this case, "-.5+L1") and press [ENTER]; the corresponding values for L2 are calculated and displayed. Repeat for L3.

This can also be accomplished with the TI-85 and TI-86. From the homescreen, press [2nd] [−] to access the LIST menu. Press F1 to obtain the { symbol, and enter the values for *i*, in order, separated by commas. Press [F2] to obtain the } symbol, then [STO▷]. Type a name for the list, perhaps I, and press [ENTER]. The list is now stored under the name I. Type the desired expression for the second list and store it under another list name, such as X. Type the desired expression for the third list and store it under an appropriate list name, such as F. (See Figure 1.)

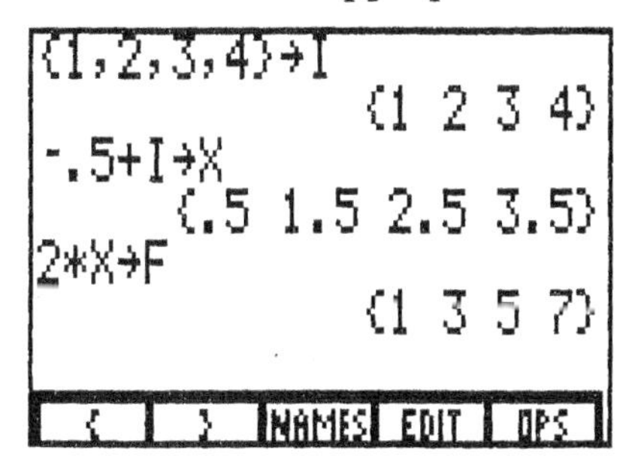

Figure 1.

Once the third list has been stored, press [F5] to access the OPS submenu, then [MORE] [F1] to obtain the sum command. Type the name of the third list and press [ENTER].

On the TI-89, these calculations can be done directly, using the Σ command in the MATH Calculus menu. From the homescreen, access this command, then type in the function (or name of the function), followed by [|], then "(x=-.5+n)". Multiply this by the size of Δx, then type a comma. Type the variable for the index, n, another comma, the first value of n, another comma, and the last value of n. Close the parentheses and press [ENTER]. For Example 1, the command should appear as follows: "Σ(2x|x=(-.5+n)*1,n,1,4)".

The Area Between Two Curves.

Approximating Definite Integrals.

Option 9 in the [MATH] menu of the TI-83 and 83/84 Plus, `fnInt`, can be used to approximate the value of a definite integral. This command can be found on the TI-85 or TI-86 by pressing [2nd] [☐] to access the CALC menu, then [F5]. The command must be followed by the function to be integrated (or the function name under which it is stored), a comma, the variable, followed by another comma, the lower limit of the integral, a comma, then the upper limit of the integral. Pressing [ENTER] executes the command. For instance, the integral in Example 4 (b) can be approximated by using the command shown in Figure 2 below.

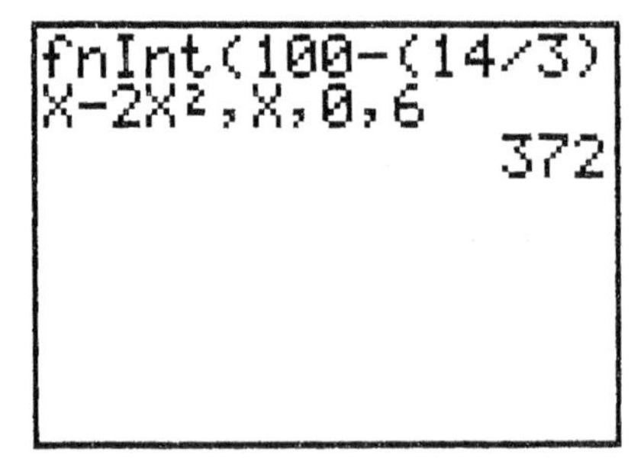

Figure 2.

If you are using the TI-89, the integral command, accessed by typing [2nd] [7], can be used to find exact values of definite integrals, as well as to evaluate indefinite integrals. To evaluate $\int 100-\frac{14}{3}x-2x^2\,dx$, access the integral command, type in the function (or name of the function), a comma, then the variable name. Press [ENTER] to see the indefinite integral. To evaluate $\int_0^6 100-\frac{14}{3}x-2x^2\,dx$, type in the following command: "∫(100-(14/3)x-2x^2,x,0,6)". Press [ENTER] to see the exact value of this definite integral, or [◆] [ENTER] for a decimal approximation.

Chapter 8 Further Techniques and Applications of Integration

LOCATION IN THE OTHER TEXTS:

Calculus with Applications, Brief Version: Chapter 8
Finite Mathematics and Calculus with Applications: Chapter 16

Integration by Parts.

Approximating Definite Integrals Graphically.

The `∫f(x)dx` command of your calculator will approximate a definite integral of a graphed function, where the upper and lower limits of integration are between `Xmin` and `Xmax`. On the TI-83 and 83/84 Plus, TI-86 and TI-89, this command also shades the region corresponding to the definite integral.

To use this command on the TI-83 or 83/84 Plus, TI-86 or TI-89, first store the function and set appropriate range variables. While viewing the graph of the function, access the `∫f(x)dx` command; it is option 7 in the `CALC` menu on the TI-83 and 83/84 Plus, it may be accessed on the TI-86 by pressing [GRAPH] [MORE] [F1] [F3]; and it is found on the TI-89 by pressing [F5] and selecting option 7. Type in the lower limit of integration, press [ENTER], then type in the upper limit and press [ENTER] again. The definite integral is approximated and the corresponding region shaded. Figure 1 below represents the results of this process, using the integral from Example 4 of this section of your text.

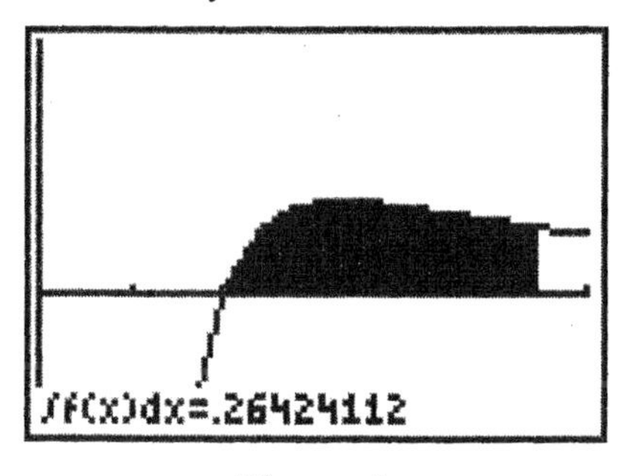

Figure 1.

If you are using the TI-85, it is most convenient to set `Xmin` and `Xmax` to be the lower and upper limits of integration, respectively. On the TI-85, access the definite integral function by pressing [GRAPH] [MORE] [F1], then [F5]. When the graph is displayed again, use the left arrow key to move the cursor as close to the lower limit of integration as possible and press [ENTER]. Use the right arrow key to move the cursor as close as possible to the upper limit and press [ENTER] once more.

Improper Integrals

Improper Integration on the TI-89

Improper integrals are set up on the TI-89 in the same manner as a definite integral. Use the ∞ symbol to enter plus or minus infinity, where necessary.

Chapter 10 Differential Equations

Solutions of Elementary and Separable Differential Equations.

Logistic Regression on the TI-83 and 83/84 Plus, TI-86 and TI-89.

Exercises 40, 47, 48, and 52 of this section of your textbook require the use of the logistic regression command of the TI-83/84 Plus, TI-86 and TI-89. The TI-85 is not equipped with this command. To solve these three exercises, enter the x-values in one list and the y-values in another. (If necessary, review page II-3 of this manual for instructions on entering lists.) Define a statistics plot to graph the two lists, turn off other functions, and set appropriate range values and press [GRAPH].

To determine the logistic function with the TI-83 and 83/84 Plus, return to the homescreen, press [STAT], and move to the `CALC` submenu by pressing the right arrow key. The Logistic regression command is option `B`. After choosing this command, type in the name of the x-list, a comma, and the name of the y-list; press [ENTER]. Because the mathematics behind the logistic regression command are complicated, the calculator takes a few seconds to complete the calculations.

To determine the logistic function with the TI-86, return to the homescreen and press [2nd] [+] [F1] to access the `STAT CALC` menu. Next, press [MORE] [F3] to access the `LgstR` command. Type in the name of the x-list, a comma, then the name of the y-list; and press [ENTER] to execute the command.

To determine the logistic function with the TI-89, from the `Data/Matrix Editor`, press [F5] and select `Logistic` as the `Calculation Type`. Proceed as described for other regression calculations on the TI-89.

Euler's Method.

Calculating y_n.

The graphing calculator method described in Example 1 of this section applies to all TI models covered by this manual. Begin by rewriting the differential equation so that the y' term is isolated on the left-hand side. Then, enter the right-hand side of the equation as Y_1 in the [Y =] menu. Return to the homescreen and store the initial values of X and Y, respectively. On a new line, type "X+", the value of *h*, press the [STO▷] key, and type "X" again. Now, *on the same line*, type a colon, followed by "Y+Y1", press [STO▷], and type "Y" again. When [ENTER] is pressed, the new value of X, x_1 is calculated and stored under the variable name X; subsequently, the new value for Y, y_1, is calculated and stored under the variable name Y. Only the new value of Y is displayed. Pressing [ENTER] a second time results in the calculation of y_2, and you may continue to press [ENTER] until the desired accuracy is obtained.

Euler's Method and the Sequence Mode of the TI-83 and 83/84 Plus.

While programs for performing Euler's method are included in Part III of this manual, it is also possible to use the sequence mode of the TI-83 to do the calculations. Press [MODE] and move the cursor to the last setting in the fourth line, `Seq`; press [ENTER]. If you now press [Y =], you will see different functions listed there than you have seen before. These functions allow you to define recursive sequences, such as those used to calculate x_i and y_i in Euler's Method.

We must now input a starting value for the subscript, `nMin`. Since the initial values are given as x_0 and y_0, type "`0`" and press [ENTER]. We will use `u(n)` to define the formula for x_{i+1}. For instance, in Example 1 of this section, $x_{i+1} = x_i + .1$; thus we will define `u(n)=u(n--1)+.1`. (The letter u is obtained by pressing [2nd] [7]; the letter n is obtained by pressing [X,T,θ,n] while in sequence mode.) Press [ENTER] to move to the next line. Here, we must enter the initial value, `u(0)`, which is the same as x_0; for Example 1, this value is 0. Simply type `0` in this case and press [ENTER]; the calculator will automatically surround this value with brackets. For `v(n)`, we will enter the corresponding formula for y_{i+1}. In Example 1, this is $y_{i+1} = y_i + (x_i - 2\ x_i\ y_i)(0.1)$. On the calculator, we type "`v(n-1)+(u(n-1)-2u(n-1)v(n-1))*.1`". (The letter `v` is obtained by pressing [2nd] [8].) Press [ENTER] and input the initial value, y_0, which is 1.5 in this case.

Press [2nd] [WINDOW] to access the `TBLSET` menu and enter "`0`" for `TblStart` and "`1`" for `ΔTbl`. Press [2nd] [GRAPH] to see the table that represents Euler's Method. (See Figure 1.)

n	u(n)	v(n)
1	.1	1.5
2	.2	1.48
3	.3	1.4408
4	.4	1.3844
5	.5	[illegible]
6	.6	1.2322
7	.7	1.1444

v(n)=1.31360384

Figure 1.

Columns two and three of this table represent the first and second columns of Table 1 in the text.

The graphs can be viewed by first setting the range variables for `Xmin`, `Xmax`, etc., to include the calculated values. You will probably want to turn the function `u(n)` "off" before pressing [GRAPH].

Chapter 11 Probability and Calculus

LOCATION IN THE OTHER TEXTS:

Finite Mathematics and Calculus with Applications: Chapter 18

Expected Value and Variance of Continuous Random Variables.

Evaluating Definite Integrals.

Recall from previous discussions that the `fnInt` function of the TI-83 or 83/84 Plus, TI-85, and TI-86, as well as the ∫ command on the TI-89, can be used to evaluate definite integrals; review the instructions on page II-40 of this manual, if necessary. The mean and variance in Example 1 of this section of your text can be approximated by using the commands shown in Figures 1 and 2.

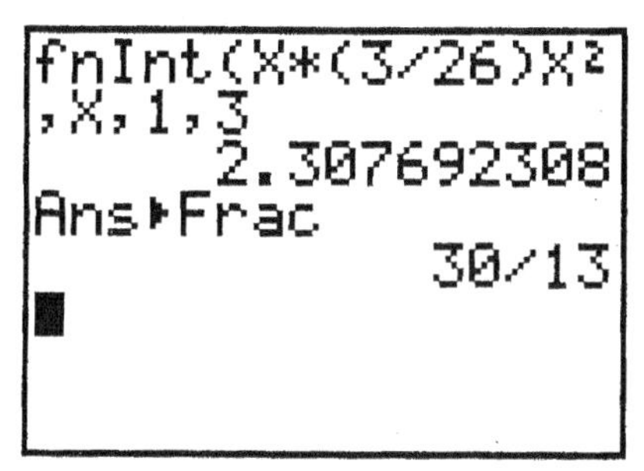

Figure 1.

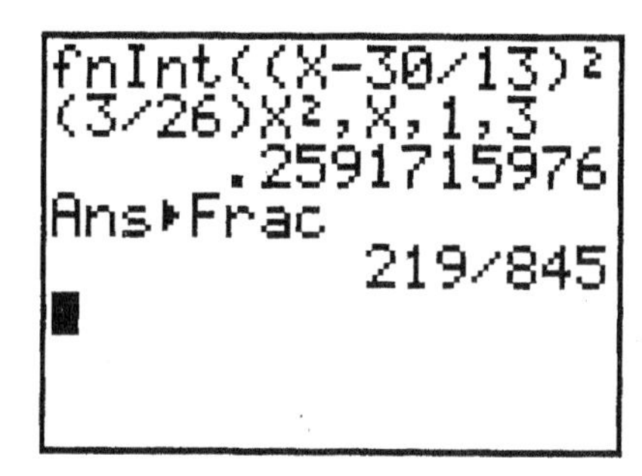

Figure 2.

The `Frac` command is the first option in the `MATH` menu of the TI-83 or 83/84 Plus. It can be found on the TI-85 and TI-86 by pressing [2nd] [×] [F5] to access the `MATH MISC` menu, then [MORE] [F1]. Accessing this command immediately after a calculation converts the answer, if possible, to a fraction.

Special Probability Density Functions.

The Normal Distribution.

Refer to page II-41 of this manual to review how to approximate definite integrals graphically with your calculator. Also, instructions for using the `normalcdf` command on the TI-83 or 83/84 Plus are located on page II-27 of this manual. Refer to them if necessary for solving Example 3 in your text.

Chapter 13 The Trigonometric Functions

Definitions of the Trigonometric Functions.

Converting Between Degree and Radian Measure.

To change the angle measure setting on your calculator, press [MODE] on the TI-83 or 83/84 Plus or TI-89, or [2nd] [MORE] on the TI-85 and TI-86. The type of measure for an angle will be shown as either `Radian` or `Degree`. Whichever is highlighted is how your calculator will treat any angles used in commands or expressions. To change the setting, move the cursor to the desired angle measure and press [ENTER].

To convert from one angle measure to another, make sure that the calculator is set in the mode to which you wish to convert your angles. In other words, if you want to convert from degrees to radians, set the calculator into `Radian` mode first; if you want to convert from radians to degrees, set the calculator into `Degree` mode first. To work Example 1 of the text, set the calculator into `Radian` mode for parts (a) and (b). To convert 45° to radians with the TI-83, from the homescreen type "`45`", then press [2nd] [MATRX] to access the `ANGLE` menu. Choose option `1`, °. The symbol will appear beside "`45`" on the homescreen; press [ENTER] to perform the conversion. On a TI 83/84 Plus, press [2nd] [APPS] to access the `ANGLE` menu.

For parts (c) and (d), first set your calculator to `Degree` mode. For part (c), from the homescreen, type "`(9π/4)`"; press [2nd] [MATRX] (or [2nd] [APPS] on a TI 83/84 Plus) and choose option `3`. The symbol ^{r} will appear on the homescreen beside the angle you entered; press [ENTER] to perform the conversion.

Similar steps are used to convert between angle measures on the TI-85 and TI-86. The ° and ^{r} symbols can be obtained by pressing [2nd] [×] to access the `MATH` menu, then [F3] for the `ANGLE` submenu. On the TI-89, these are located in the `Angle` collection of commands in the `MATH` menu.

Sine Regression on the TI-83 and 83/84 Plus, TI-86 and TI-89.

Several exercises in this chapter require the use of the sine regression feature of the TI-83 and 83/84 Plus, TI-86 and TI-89. To use this feature, first save the data to two lists, one for the x-values (in this case, the months), and one for the corresponding y-values (in this case, the average temperature). Review the instructions for entering lists on page II-3 of this manual, if necessary.

To continue with the TI-83 and 83/84 Plus, from the homescreen, press [STAT] and use the right arrow key to move to the `CALC` submenu. `SinReg` is option `C` in this menu; access it and follow it with the name of the x-list, a comma, the name of the y-list, another comma, and a function name for saving the regression equation. Press ENTER to perform the calculation.

To continue the problem with the TI-86, press [2nd] [+] F1 to access the `STAT CALC` menu, then [MORE] [F2] for the `SinR` command. Type in the name of the x-list, a comma, the name of the y-list, another comma, then a function name for saving the regression equation. Press [ENTER] to execute the command.

To complete this problem on the TI-89, from the `Data/Matrix Editor`, press [F5] and choose `SinReg` as the `Calculation Type`.

The result on either calculator is a list of coefficients, `a`, `b`, `c`, and `d`, for the function `y = a*sin(bx+c)+d` which best fits the data. You can now set up a statistics plot and graph the scatter plot and the regression equation on the same set of axes.

Part III

Programs

Introduction

This part contains hard copy of various programs for graphing calculators. The following calculators are considered:

- Texas Instruments TI-82
- Texas Instruments TI-83 and 83/84 Plus
- Texas Instruments TI-85
- Texas Instruments TI-86
- Texas Instruments TI-89
- Casio CFX-9800G
- Casio CFX-9850G
- Hewlett-Packard HP-38G

Information on how to enter these programs manually and how to run them is provided for each of these calculators at the beginning of each section. However, these short instructions do not claim to tell the whole story on how to operate the calculator; the reader is directed to the Owner's Manuals for the various calculators if more detailed information is desired.

Since the TI-82, TI-83, and TI-83/84 Plus are very similar machines, the same programs will work on either calculator; these calculators are discussed in a single section. Likewise the similarity of the TI-85 and TI-86 allows the same programs to work on either of them, and these calculators are covered in one section also. The programs for the Texas Instruments calculators are also available via the World Wide Web; see the Texas Instruments sections for more details.

The author of this part wishes to acknowledge the large amount of help he received from two people in the writing of these programs. Thomas Hungerford wrote many of the programs for the Texas Instruments calculators, while Alan Ziv provided valuable expertise and wrote several of the programs for the Casio calculators.

Programs for the TI-82, TI-83, and TI-83/84 Plus

Introduction

The following section contains programs for the TI-82, TI-83, and TI-83/84 Plus calculators. These programs allow the calculator to do various financial calculations, linear programming, numerical integration, Euler's method for the numerical solution of differential equations, and to find probabilities associated with the normal distribution.

Entering the Programs Manually: The Program Editor is used to enter a program manually. To access the Program Editor press [PGRM]. To enter a new program, select `NEW` and press [ENTER]; to edit an existing program, select `EDIT`, then move down the list of programs to the program you wish to edit. If you are creating a new program, you will first be prompted for a program name. After entering the name, you will see a colon with a cursor to its left. You are now to enter the first line of the program. Each line of a program begins with a colon; these are supplied by the Editor.

When entering a program from the Program Editor, simple commands may be entered as they appear on the keyboard. Alphabet keys are also easily accessible from the keyboard. Only upper-case letters are available. To enter a letter, press [ALPHA] then the appropriate key. Pressing [2nd] [ALPHA] locks the calculator in alphabet mode. You can easily tell when the calculator is in alphabet mode by the cursor, which flashes an `A` when in alphabet mode. The → symbol which you will need for the programs is the [STO▶] key. In addition to keyboard symbols, you will need to find the commands such as `ClrHome`, `Disp`, and `round` which appear in the programs. These commands are located in various menus; for example `ClrHome` and `Disp` are in the `I/O` submenu of the [PRGM] menu in the Program Editor. So to enter `ClrHome` from the Program Editor, you would press [PGRM], then select the `I/O` menu, then move down that menu until you reach `8:ClrHome`. Pressing [ENTER] now will place `ClrHome` at the point where you left the Program Editor. Some commands are somewhat difficult to find. On the TI-82, you will need to use the Table of Functions and Instructions in Appendix A of the Guidebook to help you find the proper menus. On the TI-83, there is another way to enter commands into the program. Pressing [2nd] [0] accesses the `CATALOG` menu. Pressing a letter key now will send you to the commands beginning with that letter. Pressing [ENTER] will place the selected command into your program. So to enter `ClrHome` by this method, you would enter the catalog, press [PRGM] to access the letter C (the calculator is already in ALPHA mode), then move down the list until the cursor points at ClrHome, then press [ENTER].

When you are finished entering the program, press [2nd] [MODE] (QUIT) to exit the Program Editor. The program is now stored under the name you have given it.

Entering the Programs from another TI-82, TI-83, or TI-83/84 Plus: Programs and other data may be transferred from a TI-82 to a TI-82 or from a TI-83 or TI-83/84 Plus to a TI-83 or TI-83/84 Plus. It is also possible to transfer files from a TI-82 to a TI-83 or a TI-83/84 Plus, although there are differences between the calculators which could lead to errors in programs transferred in this manner. To perform a transfer you need the communcation link cable which was enclosed with your calculator. Directions on how to perform this process are given in Chapter 16 of the TI-82 Guidebook and in Chapter 19 of the TI-83 Guidebook or TI-83 Plus Guidebook or TI-84-Plus Guidebook.

Entering the Programs from a Computer: The programs in this section are available on the the MyMathLab site for this book. The names of these programs are given at the beginning of each program listing. The programs may be downloaded to a PC then transferred to your calculator. You will need the TI Connect software and cable to perform this transfer. Directions on transferring programs from a PC to the calculator are included with the TI Connectivity Kit.

Running the Programs: To run a program, press [PGRM], make sure that EXEC is selected at the top of the screen, then move down the list of programs to the one you wish to run. Pressing [ENTER] will clear the screen and the name of the program will appear. Pressing [ENTER] again begins execution of the program.

Future Value of an Annuity – FVANN.82P

The following program computes the future value of an annuity given the payment amount, the rate (remember that percents should be converted to decimals), and the number of payments. Payments are made at the end of each period. On the TI-83 or TI-83/84 Plus you may also use the TVM Solver for this calculation. See Chapter 14 of the TI-83 or TI-83/84 Plus Guidebook or pages II-11 through II-16 of this manual for more details.

```
:ClrHome
:Input "PAYMENT?",R
:Input "RATE?",I
:Input "NUMBER OF PYMTS?",N
:Disp "FUTURE VALUE"
:Disp round(R*((1+I)^N-1)/I,2)
```

Present Value of an Annuity – PVANN.82P

The following program computes the present value of an annuity given the payment amount, the rate (remember that percents should be converted to decimals), and the number of payments. Payments are made at the end of each period. On the TI-83 or TI-83/84 Plus, you may also use the TVM Solver for this calculation. See Chapter 14 of the TI-83 or TI-83/84 Plus Guidebook or pages II-11 through II-16 of this manual for more details.

```
:ClrHome
:Input "PAYMENT?",R
:Input "RATE?",I
:Input "NUMBER OF PYMTS?",N
:Disp "PRESENT VALUE"
:Disp round(R*(1-(1+I)^(-N))/I,2)
```

Amortization Table – AMORT.82P

The following program asks for the following information about a loan: the starting balance, the rate per period (percents should be converted to decimals), amount of each regular payment, and the number of payments. The program displays an amortization table for this loan in the form of a matrix through which you may scroll by using the arrow keys. In this table the columns are respectively:

Payment Number	Amount of Payment	Interest for the Period	Portion to Principal	New Balance

After scrolling through the table, pressing ENTER displays the total payments made and the total interest paid. The program stores the table as matrix [A] for future viewing (until the program is run again and a new table takes its place). When the regular payment is too small, the final payment may be quite large. When the regular payment is too large, the final payment may occur before N is reached, where N is the number of payments entered at the beginning of the program. In this case the matrix will have fewer than N rows. Also note that your choice of N must be less than or equal to 99.

On the TI-83 or TI-83/84 Plus, you may also use the TVM Solver for this calculation. See Chapter 14 of the TI-83 or TI-83/84 Guidebook or pages II-11 through II-16 of this manual for more details.

```
:ClrHome
:Input "STARTING BALANCE?",B
:Input "RATE PER PERIOD?",I
:Input "PAYMENT?",P
```

```
:Input "NUMBER OF PYMTS?",N
:{N,5} → dim([A])
:1 → K
:0 → W
:Lbl 1
:If K=N+1
:Then
:Pause [A]
:ClrHome
:Disp "TOTAL PAYMENTS"
:Disp (J-1)*P+[A](J,2)
:Disp "TOTAL INTEREST"
:Disp W+[A](J,3)
:Stop
:End
:K → [A](K,1)
:P → [A](K,2)
:round(I*B,2) → [A](K,3)
:[A](K,3)+W → W
:P-[A](K,3) → [A](K,4)
:B-[A](K,4) → [A](K,5)
:[A](K,5) → B
:K+1 → K
:If K=N
:Goto 2
:If B≤P
:Goto 2
:Goto 1
:Lbl 2
:K → [A](K,1)
:B+round(I*B,2) → [A](K,2)
:round(I*B,2) → [A](K,3)
:[A](K,2) - [A](K,3) → [A](K,4)
:B-[A](K,4) → [A](K,5)
:{K,5} → dim([A])
:K → J
:N+1 → K
:Goto 1
```

Linear Programming – Maximization – SIMPLEX.82P

The following program performs the simplex method on a tableau (matrix) which has been previously input as [A]. To run the program you should store your initial tableau as [A]. When the program prompts you for the initial matrix, choose [A] from the list in the MATRX menu. The program will pause to show you intermediate matrices in the calculation, if the calculation takes more than one step. You may use the arrow keys to scroll through this matrix if it is too large for the screen. Press ENTER to continue the calculation. The final matrix likewise may be investigated by scrolling. If a solution cannot be found, the calculator reports this fact.

```
:ClrHome
:Disp "INITIAL SIMPLEX"
:Disp "MATRIX"
:Input [A]
:[A] → [B]
:dim [B] → L1
:L1(1) → R
:L1(2) → S
:seq([B](R,I),I,1,(S-1),1) → L2
:min(L2) → T
:If T ≥ 0
:Goto 1
:Lbl X
:1 → J
:Lbl 2
:If [B](R,J)=T
:Goto 3
:J+1 → J
:Goto 2
:Lbl 3
:1 → K
:Lbl 6
:If [B](K,J) ≤ 0
:Goto 4
:[[[B](K,S)/[B](K,J)]] → [E]
:K+1 → K
:Lbl 7
:If [B](K,J) ≤ 0
:Goto 5
:augment([E],[[[B](K,S)/[B](K,J)]]) → [E]
:K+1 → K
```

```
:Goto 7
:Lbl 8
:dim [E] → L3
:seq([E](1,I),I,1,L3(2),1) → L4
:min(L4) → M
:1 → I
:Lbl U
:If [B](I,J)=0
:Goto Y
:If [B](I,S)/[B](I,J)=M
:Goto V
:I+1 → I
:Goto U
:Lbl V
:If I=1
:Goto W
:*row([B](I,J)⁻¹,[B],I) → [B]
:For (K,1,I-1,1)
:*row+(⁻[B](K,J),[B],I,K) → [B]
:End
:For (K,I+1,R,1)
:*row+(⁻[B](K,J),[B],I,K) → [B]
:End
:round([B],9) → [B]
:seq([B](R,I),I,1,(S-1),1) → L2
:min(L2) → T
:If T ≥ 0
:Coto 1
:ClrHome
:Pause [B]▶Frac
:Goto X
:Lbl W
:*row([B](I,J)⁻¹,[B],1) → [B]
:For (K,2,R,1)
:*row+(⁻[B](K,J),[B],I,K) → [B]
:End
:round([B],9) → [B]
:seq([B](R,I),I,1,(S-1),1) → L2
:min(L2) → T
:If T ≥ 0
:Goto 1
```

```
:ClrHome
:Pause [B]►Frac
:Goto X
:Lbl 4
:K+1 → K
:If K>(R-1)
:Goto 9
:Goto 6
:Lbl 5
:K+1 → K
:If K>(R-1)
:Goto 8
:Goto 7
:Lbl 1
:ClrHome
:Disp "FINAL SIMPLEX"
:Disp "MATRIX"
:Disp " "
:Pause [B]►Frac
:Stop
:Lbl 9
:Disp "NO MAXIMUM SOLUTION"
:Stop
:Lbl Y
:I+1 → I
:Goto U
```

Trapezoidal Rule – TZOID.82P

The following program uses the Trapezoidal Rule to approximate the value of a definite integral. The function you wish to integrate must be stored in the variable Y_1 before you execute this program. You are also required to input the lower limit of integration (A), the upper limit of integration (B), and the number of subintervals (N).

```
:ClrHome
:Prompt A
:Prompt B
:Prompt N
:(B-A)/N → D
:0 → S
:Y1(A) → S
```

```
:Y1(B)+S → S
:For(K,1,N-1,1)
:2Y1(A+K*D)+S → S
:End
:Disp S*D/2
```

Simpson's Rule – SIMPSON.82P

The following program uses Simpson's Rule to approximate the value of a definite integral. The function you wish to integrate must be stored in the variable Y_1 before you execute this program. You are also required to input the lower limit of integration (A), the upper limit of integration (B), and the number of subintervals (N). You must choose an even number for N.

```
:ClrHome
:Prompt A
:Prompt B
:Prompt N
:(B-A)/N → D
:0 → S
:Y1(A) → S
:Y1(B)+S → S
:For(K,1,N/2,1)
:4Y1(A+(2K-1)*D)+S → S
:End
:For(K,1,N/2-1,1)
:2Y1(A+2K*D)+S → S
:End
:Disp S*D/3
```

Integration by Endpoints or Midpoint – LSUM.82P, RSUM.82P, and MSUM.82P

The following programs approximate the value of a definite integral. The function you wish to integrate must be stored in the variable Y_1 before you execute this program. You are also required to input the lower limit of integration (A), the upper limit of integration (B), and the number of subintervals (N).

Left Endpoints – LSUM.82P:

```
:ClrHome
:Prompt A
:Prompt B
:Prompt N
:(B-A)/N → D
:0 → S
:For(K,0,N-1,1)
:Y1(A+K*D)+S → S
:End
:Disp S*D
```

Right Endpoints – RSUM.82P:

```
:ClrHome
:Prompt A
:Prompt B
:Prompt N
:(B-A)/N → D
:0 → S
:For(K,1,N,1)
:Y1(A+K*D)+S → S
:End
:Disp S*D
```

Midpoints – MSUM.82P:

```
:ClrHome
:Prompt A
:Prompt B
:Prompt N
:(B-A)/N → D
:0 → S
:For(K,1,N,1)
:Y1(A+(2K-1)*D/2)+S → S
:End
:Disp S*D
```

Euler's Method – EULER.82P

The following program performs Euler's method to approximate the solution to the differential equation

$$\frac{dy}{dx} = f(x,y).$$

The function $f(x,y)$ must be stored as Y_1 before you execute this program, and each occurence of y in $f(x,y)$ should be entered as `Y`. You also must enter an initial point (both x and y coordinates), the increment between successive x values, and the value of x at which you wish to find an estimate for y.

```
:ClrHome
:Input "INITIAL X VALUE?",X
:Input "INITIAL Y VALUE?",Y
:Input "INCREMENT?",H
:Input "FINAL X VALUE?",Z
:While X≠Z
:Y+H*Y1 → Y
:X+H → X
:End
:Disp "FINAL (X,Y)"
:Disp X,Y
```

Normal Probability – NRML.82P

The following program computes the probability that a normal random variable with given mean and standard deviation lies between two given values. The probability that the random variable lies below a particular value may be computed by inputting -1E99 for the lower limit; the probability that the random variable lies above a particular value may be computed by inputting 1E99 for the upper limit. The TI-83 or TI-83/84 Plus has a built-in program called `normalcdf` which performs the work of this program. The program `normalcdf` is located in the `DISTR` menu; see the TI-83 or TI-83/84 Plus Guidebook for more details.

```
:ClrHome
:Input "MEAN?",M
:Input "STD DEV?",S
:Input "LOWER LIMIT?",A
:(A-M)/S→A
:If A<⁻10
:⁻10 → A
:If A>10
:10 → A
```

```
:Input "UPPER LIMIT?",B
:(B-M)/S→B
:If B<⁻10
:⁻10 → B
:If B>10
:10 → B
:Disp "PR(A<N(M,S)<B)="
:Disp round(fnInt(e^(X^2/⁻2),X,A,B)/√(2*π),4)
```

Programs for the TI-85 and TI-86

Introduction

The following section contains programs for the TI-85 and TI-86 calculators. These programs allow the calculator to do various financial calculations, linear programming, numerical integration, Euler's method for the numerical solution of differential equations, and to find probabilities associated with the normal distribution.

Entering the Programs Manually: The Program Editor is used to enter a program manually. To access the Program Editor press PGRM. To enter a new program, select EDIT; to edit an existing program, select EDIT, press the menu key under the name of the program you wish to edit, then press ENTER. The programs are listed in alphabetical order; if you don't see the program you want, try pressing MORE to expose the next portion of the list. If you are creating a new program, you will first be prompted for a program name. After entering the name, you will see a colon with a cursor to its left. You are now to enter the first line of the program. Each line of a program begins with a colon; these are supplied by the Editor.

When entering a program, simple commands may be entered as they appear on the keyboard. Alphabet keys are also easily accessible from the keyboard. To enter an upper case letter, press ALPHA then the appropriate key. To enter a lower case letter, press 2nd ALPHA then the appropriate key. Pressing ALPHA ALPHA locks the calculator in upper-case alphabet mode; pressing 2nd ALPHA ALPHA locks the calculator in lower-case alphabet mode. You can easily tell when the calculator is in alphabet mode by the cursor, which flashes an A when in upper-case alphabet mode and an a when in lower-case alphabet mode. The → symbol which you will need for these programs is the STO▶ key. In addition to keyboard symbols, you will need to find the commands such as ClLCD, Disp, and round which appear in the programs. These commands are located in various menus. For example, ClLCD and Disp are in the I/O menu, which is accessed by pressing F3 when entering a program. So to enter ClLCD, you would select the I/O menu by pressing F3, then move across that menu (pressing MORE to get more of the menu) until you reach ClLCD. Pressing the menu key beneath ClLCD now will place ClLCD at the point where you left the editor. Some commands are somewhat difficult to find. There is another way to enter commands into the program. Pressing 2nd CUSTOM will access the CATALOG menu. On the TI-86 the catalog will appear automatically; on the TI-86 you must additionally press F1. Pressing a letter key now will send you to the commands beginning with that letter. Pressing [ENTER] will place the selected command into your program. So to enter ClLCD by this method, you would enter the catalog, press [COS] to enter a C (the calculator is already in ALPHA mode), move down the list until the cursor points at ClLCD, then press [ENTER].

When you are finished entering the program, press [2nd] [EXIT] (QUIT) to exit the editor. The program is now stored under the name you have given it.

Entering the Programs from another TI-85 or TI-86: Programs and other data may be transferred from a TI-85 to a TI-85 or from a TI-86 to a TI-86. It is also possible to transfer programs from a TI-85 to a TI-86, although there are differences between the calculators which could lead to errors in programs transferred in this manner. To perform a transfer you need the communcation link cable which was enclosed with your calculator. Directions on how to perform this process are given in Chapter 19 of the TI-85 Guidebook and in Chapter 18 of the TI-86 Guidebook.

Entering the Programs from a Computer: The programs in this section are available on the the MyMathLab site for this book. The names of these programs are given at the beginning of each program listing. The programs may be downloaded to a PC then transferred to your calculator. You will need the TI Connectivity software and cable to perform this transfer. Directions on transferring programs from a PC to the calculator are included with the TI Connectivity Kit.

Running the Programs: To run a program, press [PGRM] then [F1] to access the NAMES submenu. Select the name of the program as you would for editing. Pressing [ENTER] will cause the name of the program to appear; pressing [ENTER] again begins execution of the program.

Future Value of an Annuity – FVANN.85P

The following program computes the future value of an annuity given the payment amount, the rate (remember that percents should be converted to decimals), and the number of payments. Payments are made at the end of each period.

```
:ClLCD
:Input "PAYMENT?",R
:Input "RATE?",I
:Input "NUMBER OF PYMTS?",N
:Disp "FUTURE VALUE"
:Disp round(R*((1+I)^N-1)/I,2)
```

Present Value of an Annuity – PVANN.85P

The following program computes the present value of an annuity given the payment amount, the rate (remember that percents should be converted to decimals), and the number of payments. Payments are made at the end of each period.

```
:ClLCD
:Input "PAYMENT?",R
:Input "RATE?",I
:Input "NUMBER OF PYMTS?",N
:Disp "PRESENT VALUE"
:Disp round(R*(1-(1+I)^(-N))/I,2)
```

Amortization Table – AMORT.85P

The following program asks for the following information about a loan: the starting balance, the rate per period (percents should be converted to decimals), amount of each regular payment, and the number of payments. The program displays an amortization table for this loan in the form of a matrix through which you can scroll by using the arrow keys. In this table the columns are respectively:

Payment Number	Amount of Payment	Interest for the Period	Portion to Principal	New Balance

After scrolling through the table, pressing ENTER displays the total payments made and the total interest paid. The program stores the table as matrix A for future viewing (until the program is run again and a new table takes its place). When the regular payment is too small, the final payment may be quite large. When the regular payment is too large, the final payment may occur before N is reached, where N is the number of payments entered at the beginning of the program. In this case the matrix will have fewer than N rows. Also note that your choice of N must be less than or equal to 255.

```
:ClLCD
:Input "STARTING BALANCE?",B
:Input "RATE PER PERIOD?",I
:Input "PAYMENT?",P
:Input "NUMBER OF PYMTS?",N
:{N,5} → dim A
:1 → K
:0 → W
:Lbl LB1
```

```
:If K==N+1
:Then
:Pause A
:ClLCD
:Disp "TOTAL PAYMENTS"
:Disp (J-1)*P+A(J,2)
:Disp "TOTAL INTEREST"
:Disp W+A(J,3)
:Stop
:End
:K → A(K,1)
:P → A(K,2)
:round(I*B,2) → A(K,3)
:A(K,3)+W → W
:P-A(K,3) → A(K,4)
:B-A(K,4) → A(K,5)
:A(K,5) → B
:K+1 → K
:If K==N
:Goto LB2
:If B≤P
:Goto LB2
:Goto LB1
:Lbl LB2
:K → A(K,1)
:B+round(I*B,2) → A(K,2)
:round(I*B,2) → A(K,3)
:A(K,2) - A(K,3) → A(K,4)
:B-A(K,4) → A(K,5)
:{K,5} → dim A
:K → J
:N+1 → K
:Goto LB1
```

Linear Programming – Maximization – SIMPLEX.85P

The following program performs the simplex method on a tableau (matrix) which has been previously input as A. To run the program you should store your initial tableau as A. When the program prompts you for the initial matrix, choose A from the list in the MATRX menu. The program will pause to show you intermediate matrices in the calculation, if the calculation takes more than one step. You may use the arrow keys to scroll through this matrix if it is too large for the screen. Press ENTER to continue the calculation. The final matrix likewise may be investigated by scrolling. If a solution cannot be found, the calculator reports this fact.

```
:ClLCD
:Disp "INITIAL SIMPLEX"
:Disp "MATRIX"
:Input A
:A → B
:dim B → L1
:L1(1) → R
:L1(2) → S
:seq(B(R,I),I,1,(S-1),1) → L2
:min(L2) → T
:If T ≥ 0
:Goto LB1
:Lbl LBX
:1 → J
:Lbl LB2
:If B(R,J)==T
:Goto LB3
:J+1 → J
:Goto LB2
:Lbl LB3
:1 → K
:Lbl LB6
:If B(K,J) ≤ 0
:Goto LB4
:[[B(K,S)/B(K,J)]] → E
:K+1 → K
:Lbl LB7
:If B(K,J) ≤ 0
:Goto LB5
:aug(E,[B(K,S)/B(K,J)]) → E
:K+1 → K
:Goto LB7
:Lbl LB8
:dim E → L3
:seq(E(1,I),I,1,L3(2),1) → L4
:min(L4) → M
```

```
:1 → I
:Lbl LBU
:If B(I,J)==0
:Goto LBY
:If B(I,S)/B(I,J)==M
:Goto LBV
:I+1 → I
:Goto LBU
:Lbl LBV
:If I==1
:Goto LBW
:multR(B(I,J)⁻¹,B,I) → B
:For (K,1,I-1,1)
:mRAdd(⁻B(K,J),B,I,K) → B
:End
:For (K,I+1,R,1)
:mRAdd(⁻B(K,J),B,I,K) → B
:End
:round(B,11) → B
:seq(B(R,I),I,1,(S-1),1) → L2
:min(L2) → T
:If T ≥ 0
:Goto LB1
:ClLCD
:Pause B▶Frac
:Goto LBX
:Lbl LBW
:multR(B(I,J)⁻¹,B,1) → B
:For (K,2,R,1)
:mRAdd(⁻B(K,J),B,I,K) → B
:End
:round(B,11) → B
:seq(B(R,I),I,1,(S-1),1) → L2
:min(L2) → T
:If T ≥ 0
:Goto LB1
:ClLCD
:Pause B▶Frac
:Goto LBX
:Lbl LB4
:K+1 → K
```

```
:If K>(R-1)
:Goto LB9
:Goto LB6
:Lbl LB5
:K+1 → K
:If K>(R-1)
:Goto LB8
:Goto LB7
:Lbl LB1
:ClLCD
:Disp "FINAL SIMPLEX"
:Disp "MATRIX"
:Disp " "
:Pause B▶Frac
:Stop
:Lbl LB9
:Disp "NO MAXIMUM SOLUTION"
:Stop
:Lbl LBY
:I+1 → I
:Goto LBU
```

Trapezoidal Rule – TZOID.85P

The following program uses the Trapezoidal Rule to approximate the value of a definite integral. The function you wish to integrate must be stored in the variable y_1 before you execute this program. You are also required to input the lower limit of integration (A), the upper limit of integration (B), and the number of subintervals (N).

```
:ClLCD
:Prompt A
:Prompt B
:Prompt N
:(B-A)/N → D
:0 → S
:A → x
:y1 → S
:B → x
:y1+S → S
:For(K,1,N-1,1)
```

```
:A+K*D → x
:2y1+S → S
:End
:Disp S*D/2
```

Simpson's Rule – SIMPSON.85P

The following program uses Simpson's Rule to approximate the value of a definite integral. The function you wish to integrate must be stored in the variable y_1 before you execute this program. You are also required to input the lower limit of integration (A), the upper limit of integration (B), and the number of subintervals (N). You must choose an even number for N.

```
:ClLCD
:Prompt A
:Prompt B
:Prompt N
:(B-A)/N → D
:0 → S
:A → x
:y1 → S
:B → x
:y1+S → S
:For(K,1,N/2,1)
:A+(2K-1)*D → x
:4y1+S → S
:End
:For(K,1,N/2-1,1)
:A+2K*D → x
:2y1+S → S
:End
:Disp S*D/3
```

Integration by Endpoints or Midpoint – LSUM.85P, RSUM.85P, and MSUM.85P

The following programs approximate the value of a definite integral. The function you wish to integrate must be stored in the variable y_1 before you execute this program. You are also required to input the lower limit of integration (A), the upper limit of integration (B), and the number of subintervals (N).

Left Endpoints – LSUM.85P:

```
:ClLCD
:Prompt A
:Prompt B
:Prompt N
:(B-A)/N → D
:0 → S
:For(K,0,N-1,1)
:A+K*D → x
:y1+S → S
:End
:Disp S*D
```

Right Endpoints – RSUM.85P:

```
:ClLCD
:Prompt A
:Prompt B
:Prompt N
:(B-A)/N → D
:0 → S
:For(K,1,N,1)
:A+K*D → x
:y1+S → S
:End
:Disp S*D
```

Midpoints – MSUM.85P:

```
:ClLCD
:Prompt A
:Prompt B
:Prompt N
:(B-A)/N → D
:0 → S
:For(K,1,N,1)
:A+(2K-1)*D/2 → x
:y1+S → S
:End
:Disp S*D
```

Euler's Method – EULER1.85P

The following program performs Euler's method to approximate the solution to the differential equation

$$\frac{dy}{dx} = f(x,y).$$

The function $f(x)$ must be stored as y_1 before you execute this program, and each occurence of y in $f(x,y)$ should be entered as Y. You also must enter an initial point (both x and y coordinates), the increment between successive x values, and the value of x at which you wish to find an estimate for y. The TI-86 has a program which graphs an approximate solution for a differential equation using Euler's method; see Chapter 10 of the TI-86 Guidebook for more details.

```
:ClLCD
:Input "INITIAL X VALUE?",x
:Input "INITIAL Y VALUE?",Y
:Input "INCREMENT?",H
:Input "FINAL X VALUE?",Z
:While x≠Z
:Y+H*y1 → Y
:x+H → x
:End
:Disp "FINAL (X,Y)"
:Disp x,Y
```

Normal Probability – NRML.85P

The following program computes the probability that a normal random variable with given mean and standard deviation lies between two given values. The probability that the random variable lies below a particular value may be computed by inputting -1E99 for the lower limit; the probability that the random variable lies above a particular value may be computed by inputting 1E99 for the upper limit.

```
:ClLCD
:Input "MEAN?",M
:Input "STD DEV?",S
:Input "LOWER LIMIT?",A
:(A-M)/S→A
:If A<⁻10
:⁻10 → A
:If A>10
:10 → A
```

```
:Input "UPPER LIMIT?",B
:(B-M)/S→B
:If B<⁻10
:⁻10 → B
:If B>10
:10 → B
:Disp "PR(A<N(M,S)<B)="
:Disp round(fnInt(e^(X^2/⁻2),X,A,B)/√(2*π),4)
```

Programs for the TI-89

Introduction

The following section contains programs for the TI-89 calculator. These programs allow the calculator to do various financial calculations, linear programming, numerical integration, Euler's method for the numerical solution of differential equations, and to find probabilities associated with the normal distribution.

Entering the Programs Manually: The Program Editor is used to enter a program manually. To access the Program Editor press APPS 6. To enter a new program, select 3:New. You will first be prompted for a program name and for the name of the folder in which you wish to store the program. After entering the name, you will see the name of the program, the commands Pgrm and EndPgrm, and a colon with a blank line following it. You are to enter the first line of the program on that blank line. Each line of a program begins with a colon; these are supplied by the Editor. When you are done, pressing 2nd ESC (QUIT) or HOME returns you to the home screen.The program is now stored under the name you have given it.

When entering a program, simple commands may be entered as they appear on the keyboard. Alphabet keys are also easily accessible from the keyboard. To enter an lower case letter, press alpha then the appropriate key. To enter a upper case letter, press ⇑ ALPHA then the appropriate key. Pressing alpha alpha or a-lock (2nd alpha) locks the calculator in alphabet mode. You can easily tell when the calculator is in alphabet mode by the icon a which appears in the status line at the bottom of the calculator screen. The → symbol which you will need for these programs is the STO▶ key. In addition to keyboard symbols, you will need to find the commands such as ClrIO, Disp, and round which appear in the programs. These commands are located in various menus. For example, round is in the MATH Number menu, which is accessed by by pressing 2nd 5, then selecting 1:Number. Pressing the right arrow key exposes a long list of commands. Moving down the list to 3:round(and pressing ENTER will place round(at the point where you left the Program Editor. The menus for some commands are somewhat difficult to find; all commands are listed in the catalog, which is accessed by pressing CATALOG. Pressing a letter key now will send you to the commands beginning with that letter, and pressing ENTER will place the selected command into your program. So to enter ClrIO by this method, you would enter the catalog, press) to enter a c (the calculator is already in alphabet mode), move down the list until the cursor points at `ClrIO`, then press ENTER.

To edit an existing program, access the Program Editor and select `2:Open`. With the Variable box selected, press the right arrow key to see a list of programs in the current folder. Move the cursor down to the name of the program you wish to edit, then press ENTER twice. The programs are listed in alphabetical order; if you don't see the program you want, continue using the down arrow to expose more of the list.

Entering the Programs from another TI-89: Programs and other data may be transferred from a TI-89 to another TI-89. To perform a transfer you need the communication link cable which was enclosed with your calculator. Directions on how to perform this process are given in Chapter 22 of the TI-89 Guidebook.

Entering the Programs from a Computer: The programs in this section are available on the the MyMathLab site for this book. The names of these programs are given at the beginning of each program listing. The programs may be downloaded to your computer and then transferred to your calculator. You will need the TI Connect software and cable to perform this transfer. Directions on transferring programs from a PC to the calculator are included with TI Connectivity Kit.

Running the Programs: To run a program, enter the name of the program on the entry line of the Home screen. Pressing ENTER begins execution of the program. For example, entering `fvann()` on the entry line, then pressing ENTER would start the first program listed below. You can also locate the program using the VAR-LINK feature on the calculator (2nd −) – see Chapter 21 of the TI-89 Guidebook for details.

Future Value of an Annuity – FVANN.89P

The following program computes the future value of an annuity given the payment amount, the rate (remember that percents should be converted to decimals), and the number of payments. Payments are made at the end of each period.

```
:fvann()
:Prgm
:ClrIO
:Input "Payment?",r
:Input "Rate?",i
:Input "Number of payments?",n
:Disp "Future value"
:Disp round(r*((1+i)^n-1)/i,2)
:EndPrgm
```

Present Value of an Annuity – PVANN.89P

The following program computes the present value of an annuity given the payment amount, the rate (remember that percents should be converted to decimals), and the number of payments. Payments are made at the end of each period.

```
:pvann()
:Prgm
:ClrIO
:Input "Payment?",r
:Input "Rate?",i
:Input "Number of payments?",n
:Disp "Present value"
:Disp round(r*(1-(1+i)^(-n))/i,2)
:EndPrgm
```

Amortization Table – AMORT.89P

The following program asks for the following information about a loan: the starting balance, the rate per period (percents should be converted to decimals), amount of each regular payment, and the number of payments. The program creates an amortization table for this loan, and displays the total payments made and the total interest paid over the life of the loan. In the amortization table the columns are respectively:

Payment Number	Amount of Payment	Interest for the Period	Portion to Principal	New Balance

The program stores the amortization table as matrix a for future viewing (until the program is run again and a new table takes its place). To view the table, open the Data/Matrix Editor (APPS 6) and select `1:Current`. When the regular payment is too small, the final payment may be quite large. When the regular payment is too large, the final payment may occur before n is reached, where n is the number of payments entered at the beginning of the program. In this case the matrix will have fewer than n rows. Also note that your choice of n must be less than or equal to 999.

```
:amort()
:Prgm
:ClrIO
:Input "Starting balance?",b
:Input "Rate per period?",i
:Input "Payment?",p
```

```
:Input "Number of payments?",n
:1 → k
:0 → w
:{} → col1
:{} → col2
:{} → col3
:{} → col4
:{} → col5
:Lbl lbl1
:If k=n+1 Then
:NewData a,col1,col2,col3,col4,col5
:ClrIO
:Disp "Total Payments"
:Disp (j-1)*p+col2[j]
:Disp "Total Interest"
:Disp w+col3[j]
:Stop
:EndIf
:k → col1[k]
:p → col2[k]
:round(i*b,2) → col3[k]
:col3[k]+w → w
:p-col3[k] → col4[k]
:b-col4[k] → col5[k]
:col5[k] → b
:k+1 → k
:If k=n
:Goto lbl2
:If b≤p
:Goto lbl2
:Goto lbl1
:Lbl lbl2
:k → col1[k]
:b+round(i*b,2) → col2[k]
:round(i*b,2) → col3[k]
:col2[k]-col3[k] → col4[k]
:b-col4[k] → col5[k]
:k → j
:n+1 → k
:Goto lbl1
:EndPrgm
```

Linear Programming – Maximization – SIMPLEX.89P

The following program performs the simplex method on a tableau (matrix) which you provide. When the program prompts you for the initial matrix, you may input it at the prompt. If you have previously stored the matrix, simply enter the variable name at the prompt. The program will pause to show you intermediate matrices in the calculation, if the calculation takes more than one step. You may use the cursor pad to scroll through this matrix if it is too large for the screen. Press ENTER to continue the calculation. The final matrix likewise may be investigated by scrolling. If a solution cannot be found, the calculator reports this fact.

```
:simplex()
:Prgm
:ClrIO
:Disp "Initial simplex matrix"
:Input a
:a → b
:dim(b) → l1
:l1[1] → r
:l1[2] → s
:seq(b[r,i],i,1,s-1,1) → l2
:min(l2) → t
:If t ≥ 0
:Goto lbl1
:Lbl lbl13
:1 → j
:Lbl lbl2
:If b[r,j]=t
:Goto lbl3
:j+1 → j
:Goto lbl2
:Lbl lbl3
:1 → k
:Lbl lbl6
:If b[k,j] ≤ 0
:Goto lbl4
:[[b[k,s]/(b[k,j])]] → e
:k+1 → k
:Lbl lbl7
:If b[k,j] ≤ 0
```

```
:Goto lbl5
:augment(e,[[b[k,s]/(b[k,j])]]) → e
:k+1 → k
:Goto lbl7
:Lbl lbl8
:dim(e) → l3
:seq(e[1,i],i,1,l3[2],1) → l4
:min(l4) → M
:1 → i
:Lbl lbl10
:If b[i,j]=0
:Goto lbl14
:If b[i,s]/(b[i,j])=m
:Goto lbl11
:i+1 → i
:Goto lbl10
:Lbl lbl11
:If i=1
:Goto lbl12
:mRow(b[i,j]^(⁻1),b,i) → b
:For k,1,i-1,1
:mRowAdd(⁻b[k,j],b,i,k) → b
:EndFor
:For k,i+1,r,1
:mRowAdd(⁻b[k,j],b,i,k) → b
:EndFor
:seq(b[r,i],i,1,s-1,1) → l2
:min(l2) → t
:If t ≥ 0
:Goto lbl1
:ClrIO
:Pause b
:Goto lbl13
:Lbl lbl12
:mRow(b[i,j]^(⁻1),b,1) → b
:For k,2,r,1
:mRowAdd(⁻b[k,j],b,i,k) → b
:EndFor
:seq(b[r,i],i,1,s-1,1) → l2
:min(l2) → t
:If t ≥ 0
```

```
:Goto lbl1
:ClrIO
:Pause b
:Goto lbl13
:Lbl lbl4
:k+1 → k
:If k>r-1
:Goto lbl9
:Goto lbl6
:Lbl lbl5
:k+1 → k
:If k>r-1
:Goto lbl8
:Goto lbl7
:Lbl lbl1
:ClrIO
:Disp "Final Simplex Matrix"
:Disp " "
:Pause b
:Stop
:Lbl lbl9
:Disp "No Maximum Solution"
:Stop
:Lbl lbl14
:i+1 → i
:Goto lbl10
:EndPrgm
```

Trapezoidal Rule – TZOID.89P

The following program uses the Trapezoidal Rule to approximate the value of a definite integral. The function you wish to integrate must be stored in the variable y_1 before you execute this program. You are also required to input the lower limit of integration (a), the upper limit of integration (b), and the number of subintervals (n).

```
:tzoid()
:Prgm
:ClrIO
:Prompt a
:Prompt b
```

```
:Prompt n
:(b-a)/n → d
:0 → s
:a → x
:y1(x) → s
:b → x
:y1(x)+s → s
:For k,1,n-1,1
:a+k*d → x
:2*y1(x)+s → s
:EndFor
:Disp s*d/(2.)
:EndPrgm
```

Simpson's Rule – SIMPSON.89P

The following program uses Simpson's Rule to approximate the value of a definite integral. The function you wish to integrate must be stored in the variable y_1 before you execute this program. You are also required to input the lower limit of integration (a), the upper limit of integration (b), and the number of subintervals (n). You must choose an even number for n.

```
:simpson()
:Prgm
:ClrIO
:Prompt a
:Prompt b
:Prompt n
:(b-a)/n → d
:0 → s
:a → x
:y1(x) → s
:b → x
:y1(x)+s → s
:For k,1,n/2,1
:a+(2*k-1)*d → x
:4*y1(x)+s → s
:EndFor
:For k,1,n/2-1,1
:a+2*k*d → x
:2*y1(x)+s → s
```

```
:EndFor
:Disp s*d/(3.)
:EndPrgm
```

Integration by Endpoints or Midpoint – LSUM.89P, RSUM.89P, MSUM.89P

The following programs approximate the value of a definite integral. The function you wish to integrate must be stored in the variable y_1 before you execute this program. You are also required to input the lower limit of integration (a), the upper limit of integration (b), and the number of subintervals (n).

Left Endpoints – LSUM.89P:

```
:lsum()
:Prgm
:ClrIO
:Prompt a
:Prompt b
:Prompt n
:(b-a)/n → d
:0 → s
:For k,0,n-1,1
:a+k*d → x
:y1(x)+s → s
:EndFor
:Disp s*d
:EndPrgm
```

Right Endpoints – RSUM.89P:

```
:rsum()
:Prgm
:ClrIO
:Prompt a
:Prompt b
:Prompt n
:(b-a)/n → d
:0 → s
:For k,1,n,1
```

```
:a+k*d → x
:y1(x)+s → s
:EndFor
:Disp s*d
:EndPrgm
```

Midpoints – MSUM.89P:

```
:msum()
:Prgm
:ClrIO
:Prompt a
:Prompt b
:Prompt n
:(b-a)/n → d
:0 → s
:For k,1,n,1
:a+(2*k-1)*d/2 → x
:y1(x)+s → s
:EndFor
:Disp s*d
:EndPrgm
```

Euler's Method – EULER.89P

The following program performs Euler's method to approximate the solution to the differential equation

$$\frac{dy}{dx} = f(x,y).$$

The function $f(x,y)$ must be stored as y_1 before you execute this program, and each occurence of y in $f(x,y)$ should be entered as y. You also must enter an initial point (both x and y coordinates), the increment between successive x values, and the value of x at which you wish to find an estimate for y.

```
:euler()
:Prgm
:ClrIO
:Input "Initial x value?",x
:Input "Initial y value?",y
:Input "Increment?",h
```

```
:Input "Final x value?",z
:While x≠z
:y+h*y1(x) → y
:x+h → x
:EndWhile
:Disp "Final (x,y)"
:Disp x,y
:EndPrgm
```

Normal Probability – NRML.89P

The following program computes the probability that a normal random variable with given mean and standard deviation lies between two given values. The probability that the random variable lies below a particular value may be computed by inputting $-\infty$ for the lower limit; the probability that the random variable lies above a particular value may be computed by inputting ∞ for the upper limit. The infinity sign can be found on the keyboard as [♦] [CATALOG].

```
:nrml()
:Prgm
:ClrIO
:Input "Mean?",m
:Input "Standard deviation?",s
:Input "Lower limit?",a
:(a-m)/s→a
:Input "Upper limit?",b
:(b-m)/s→b
:Disp "Pr(a<N(m,s)<b)="
:Disp nInt(e^(⁻x^2/2),x,a,b)/√(2*π))
:EndPrgm
```

Programs for the Casio CFX-9800G

Introduction

The following section contains programs for the Casio CFX-9800G calculator. These programs allow the calculator to do various financial calculations, numerical intergration, Euler's method for the numerical solution of differential equations, and to find probabilities associated with the normal distribution.

Entering the Programs Manually: You should enter these programs using the File Editor Mode, which allows you to choose a name for each program. To enter File Editor Mode, select the `PGRM` icon from the main menu, then press `F2`, which is labelled `EDT`. Selecting `F1` (labelled `NEW`) at this point will allow you to begin entering a program. You are first prompted for a filename; after inputting the filename press `ENTER`; you may then begin entering the program itself.

The program is entered by typing each line onto the calculator screen; to enter variable names and other data, a set of alphabet keys is provided. The `ALPHA` key activates the red versions of the keys, which are the letters of the alphabet and other symbols; `SHIFT` `ALPHA` locks the calculator in alphabet entry mode. Special symbols and programming commands are also needed. These may be placed into your program in three ways. First, the keyboard may be used to input the symbols marked on the keys. Second, pressing `F6` (labelled `SYM`) in the initial editor gives access to other symbols. Third, programming symbols such as ? or ⇒ may be entered by pressing `SHIFT` `PRGM`, then accessing the menus contained therein. Each line of the program is ended with the `EXE` key; the Editor will show a carriage return symbol at this point. These carriage returns are not shown in the program listings that follow.

Entering the Programs from another CFX-9800G: It is possible to download programs and other data from one CFX-9800G to another. The Casio SB-62 cable is used for these transfers. For details on how to perform such transfers, consult Chapter 13 in the Owner's Manual.

Running the Programs: To run the program while in File Editor Mode, simply select the name of the program from the list which appears when you enter the Mode. Pressing `F4` (or `RUN`) will start the program. If you are not in File Editor Mode, the program may be run by pressing `SHIFT` `PGRM` `F3` (labelled `Prg`). The calculator will ask for the program's name. The name must be entered enclosed by double quotes; these are accessed by pressing [SHIFT] [ALPHA] [F2]. After entering the name, pressing [EXE] runs the program.

Future Value of an Annuity

The following program computes the future value of an annuity given the payment amount, the rate (remember that percents should be converted to decimals), and the number of payments. Payments are made at the end of each period.

```
"PAYMENT"?  → R
"RATE"?  → I
"NO. OF PAYMENTS"?  → N
"FUTURE VALUE"
Fix 2
R×((1+I)^N-1)÷I
```

Present Value of an Annuity

The following program computes the present value of an annuity given the payment amount, the rate (remember that percents should be converted to decimals), and the number of payments. Payments are made at the end of each period.

```
"PAYMENT"?  → R
"RATE"?  → I
"NO. OF PAYMENTS"?  → N
"PRESENT VALUE"
Fix 2
R×(1-(1+I)^(-N))÷I
```

Payment to Amortize a Loan

The following program computes the payment needed to amortize a loan given the amount of the loan, the rate (remember that percents should be converted to decimals), and the number of payments.

```
"AMOUNT OF LOAN"?  → P
"RATE"?  → I
"NO. OF PAYMENTS"?  → N
```

```
"PYMT TO AMORTIZE"
Fix 2
P×I÷(1-(1+I)^(-N))
```

Trapezoidal Rule

The following program uses the Trapezoidal Rule to approximate the value of a definite integral. The function you wish to integrate must be stored in the Function Memory as f_6 before you execute this program. You are also required to input the lower limit of integration (A), the upper limit of integration (B), and the number of subintervals (N).

```
"A="?→A
"B="?→B
"N="?→N
(B-A)÷N→D
N→K
0→S
A→X
Lbl 0
S+f6D→S
X+D→X
Dsz K
Goto 0
A→X
f6 →E
B→X
f6 →F
S+(F-E)D÷2→S
S
```

Simpson's Rule

The following program uses Simpson's Rule to approximate the value of a definite integral. The function you wish to integrate must be stored in the variable f_6 before you execute this program. You are also required to input the lower limit of integration (A), the upper limit of integration (B), and the number of subintervals (N). You must choose an even number for N. You can also use the calculator's built-in numerical integration program, which uses Simpson's Rule as its algorithm. See the Owner's Manual for more information on the built-in function.

```
"A="?→A
"B="?→B
"N="?→N
(B-A)÷N → D
0 → S
A → X
f6 → S
B → X
f6+S → S
N÷2 → K
Lbl 0
A+(2K-1)D → X
S+4f6 → S
Dsz K
Goto 0
(N÷2)-1 → K
Lbl 1
A+2KD → X
S+2f6 → S
Dsz K
Goto 1
SD÷3
```

Integration by Endpoints or Midpoint

The following programs approximate the value of a definite integral. The function you wish to integrate must be stored in the Function Memory as f_6 before you execute this program. You are also required to input the lower limit of integration (A), the upper limit of integration (B), and the number of subintervals (N).

Left Endpoints:

```
"A="?→A
"B="?→B
"N="?→N
(B-A)÷N→D
N→K
0→S
A→X
Lbl 0
```

```
S+f6D→S
X+D→X
Dsz K
Goto 0
S
```

Right Endpoints:

```
"A="?→A
"B="?→B
"N="?→N
(B-A)÷N→D
N→K
0→S
A→X
X+D→X
Lbl 0
S+f6D→S
X+D→X
Dsz K
Goto 0
S
```

Midpoints:

```
"A="?→A
"B="?→B
"N-"?→N
(B-A)÷N→D
N→K
0→S
A→X
X+.5D→X
Lbl 0
S+f6D→S
X+D→X
Dsz K
Goto 0
S
```

Euler's Method

The following program performs Euler's method to approximate the solution to the differential equation

$$\frac{dy}{dx} = f(x).$$

The function $f(x)$ must be stored as f_6 before you execute this program. You also must enter an initial point (both x and y coordinates), the increment between successive x values, and the value of x at which you wish to find an estimate for y.

```
"INITIAL X VALUE"?→X
"INITIAL Y VALUE"?→Y
"INCREMENT"?→H
"FINAL X VALUE"?→Z
Lbl 1
Y+H×f6 → Y
X+H → X
X≠Z ⇒ Goto 1
"FINAL X VALUE"
X ▲
"FINAL Y VALUE"
Y
```

Normal Probability

The following program computes the probability that a normal random variable with given mean and standard deviation lies between two given values. The probability that the random variable lies below a particular value may be computed by inputting -1E99 for the lower limit; the probability that the random variable lies above a particular value may be computed by inputting 1E99 for the upper limit.

```
"MEAN"?→M
"STD DEV"?→S
"LOWER LIMIT"?→A
"UPPER LIMIT"?→B
(A-M)÷S→A
(B-M)÷S→B
Lbl 1
A<-10 ⇒ Goto 2
A>10 ⇒ Goto 3
Lbl 4
B<-10 ⇒ Goto 5
```

```
B>10 ⇒ Goto 6
"PR(A<N(M,S)<B)="
Fix 4
∫(e(X^2÷-2),A,B)÷√(2×π) ▲
Lbl 2
-10 → A
Goto 1
Lbl 3
10 → A
Goto 1
Lbl 5
-10 → B
Goto 4
Lbl 6
10 → B
Goto 4
```

Programs for the Casio CFX-9850G

Introduction

The following section contains programs for the Casio CFX-9850G calculator. These programs allow the calculator to do various financial calculations, linear programming, numerical intergration, Euler's method for the numerical solution of differential equations, and to find probabilities associated with the normal distribution.

Entering the Programs Manually: To enter a program manually, select the `PGRM` icon from the main menu, then press `F3`, which is labelled `NEW`. You are first prompted for a program name; after inputting this name and pressing `ENTER`, you may begin entering the program itself.

The `ALPHA` key activates the red versions of the keys, which are the letters of the alphabet and other symbols; `SHIFT` `ALPHA` locks the calculator in alphabet entry mode. Special symbols and programming commands are also needed. These may be placed into your program in three ways. First, the keyboard may be used to input the symbols marked on the keys. Second, pressing `F6` (`SYBL`) in the initial editor gives access to other symbols. Third, programming symbols such as ? or ⇒ may be entered by pressing `SHIFT` `PRGM`, then accessing the menus contained therein. Each line of the program is ended with the `EXE` key; the Editor will show a carriage return symbol at this point. These carriage returns are not shown in the program listings that follow.

Entering the Programs from another CFX-9850G: It is possible to download programs and other data from one CFX-9850G to another. The Casio SB-62 cable is used for these transfers. For details on how to perform such transfers, consult Chapter 20 in the Owner's Manual.

Running the Programs: To run the program, first select the `PGRM` icon from the main menu, then select the name of the program from the list which appears. Pressing `F1` (labelled `EXE` on the calculator screen) will start the program.

Future Value of an Annuity

The following program computes the future value of an annuity given the payment amount, the rate (remember that percents should be converted to decimals), and the number of payments. Payments are made at the end of each period.

```
"PAYMENT"?  → R
"RATE"?  → I
"NO. OF PAYMENTS"?  → N
"FUTURE VALUE"
Fix 2
R×((1+I)^N-1)÷I ▲
ClrText
Norm
```

Present Value of an Annuity

The following program computes the present value of an annuity given the payment amount, the rate (remember that percents should be converted to decimals), and the number of payments. Payments are made at the end of each period.

```
"PAYMENT"?  → R
"RATE"?  → I
"NO. OF PAYMENTS"?  → N
"PRESENT VALUE"
Fix 2
R×(1-(1+I)^(-N))÷I ▲
ClrText
Norm
```

Amortization Table

The following program asks for the following information about a loan: the starting balance, the rate per period (percents should be converted to decimals), amount of each regular payment, and the number of payments. The program displays an amortization table for this loan in the form of a matrix through which you can scroll by using the arrow keys. In this table the columns are respectively:

Payment Number	Amount of Payment	Interest for the Period	Portion to Principal	New Balance

After scrolling through the table, pressing ENTER displays the total payments made and the total interest paid. The program stores the table as matrix `Mat A` for future viewing (until the program is run again and a new table takes its place). When the regular payment is too small, the final payment may be quite large. When the regular payment is too large, the final payment may occur before `N` is reached, where `N` is the number of payments entered at the beginning of the program. In this case the matrix will have rows of zeroes filling the matrix below the line of the final payment.

Important Notes: Before running this program you must open the Matrix Editor and change the dimensions of `Mat A` to `N` by 5, where `N` is again the number of payments. To do this, select the MAT icon from the initial menu, note that the row labelled `Mat A` is highlighted, then type whatever value `N` has, then EXE, then 5, then EXE again. Pressing the MENU key now will return you to the initial menu screen. Also note that your choice of `N` must always be less than or equal to 255, and that the amount of available memory on your calculator greatly affects how large `N` may be before causing the memory to overflow.

```
Fill(0, Mat A)
Fix 2
"STARTING BALANCE"?→B
"RATE PER PERIOD"?→I
"PAYMENT"?→P
"NUMBER OF PAYMENTS"?→N
1→K
0→W
Lbl 1
If K=N+1
Then Mat A ▲
ClrText
"TOTAL PAYMENTS"
(J-1)×P+Mat A[J,2] ▲
"TOTAL INTEREST"
W+Mat A[J,3]
Norm
Stop
IfEnd
K→Mat A[K,1]
P→Mat A[K,2]
```

```
I×B
Rnd
Ans→Mat A[K,3]
Mat A[K,3]+W→W
P-Mat A[K,3]→Mat A[K,4]
B-Mat A[K,4]→Mat A[K,5]
Mat A[K,5]→ B
K+1→K
K=N⇒Goto 2
B≤P⇒Goto 2
Goto 1
Lbl 2
K→Mat A[K,1]
I×B
Rnd
Ans→S
B+S→Mat A[K,2]
S→Mat A[K,3]
Mat A[K,2] - Mat A[K,3] → Mat A[K,4]
B-Mat A[K,4]→Mat A[K,5]
K→J
N+1→K
Goto 1
```

Linear Programming – Maximization

The following program performs the simplex method on a tableau (matrix) which has been previously input as `Mat A`. To run the program you should store your initial tableau as `Mat A`. The program stores the final matrix as `Mat F` and displays it on the screen; you may use the arrow keys to scroll through this matrix.

In this program there are two commands which are rather hard to locate: the `Mat→List` command is not entered using the → key on the keyboard, but is instead entered by pressing [OPTN] [F2] [F2]. The ⌟ command in the seventh line from the end of the program is entered by pressing the [a^b/c] key on the keyboard.

```
"INITIAL SIMPLEX MATRIX IN Mat A"
Dim Mat A
List Ans[1] → R
List Ans[2] → C
```

```
Mat A → Mat F
Do
0 → P
Trn Mat F → Mat T
For 1 → K To C-1
Mat F[R,K]<0 ⇒2→P
Next
For 1 → K To R-1
Mat F[K,C]<0⇒1→P
Next
If P=1
Then Mat→List(Mat F,C)→List 6
0 → List 6[R]
Min(List 6) → M
For 1 → K To R-1
M=List 6[K] ⇒ K → I
Next
Mat→List(Mat T,I) → List 6
0 → List 6[C]
Min(List 6) → M
If M<0
Then For 1 → K To C-1
M=List 6[K] ⇒ K → J
Next
Else 0 → P
IfEnd
IfEnd
If P=2
Then Mat→List(Mat T,R)→List 6
0 → List 6[C]
Min(List 6) → M
For 1 → K To C-1
M=List 6[K] ⇒ K → J
Next
Mat→List(Mat F,J)→List 6
0 → List 6[R]
For 1→K To R-1
List 6[K]>0 ⇒ List 6[K]÷Mat F[K,C]→List 6[K]
List 6[K]≤0 ⇒ 0 →List 6[K]
Next
Max(List 6)→M
```

```
If M>0
Then For 1 → K To R-1
M=List 6[K] ⇒ K → I
Next
Else 0 → P
IfEnd
IfEnd
If P≠0
Then *Row (1 ⌟ (Mat F[I,J])),F,I
For 1 → K To R
K ≠ I ⇒ *Row+ (-Mat F[K,J]),F,I,K
Next
IfEnd
LpWhile P≠0
Mat F
```

Trapezoidal Rule

The following program uses the Trapezoidal Rule to approximate the value of a definite integral. The function you wish to integrate must be stored in the Function Memory as f_6 before you execute this program. You are also required to input the lower limit of integration (A), the upper limit of integration (B), and the number of subintervals (N).

```
"A="?→A
"B="?→B
"N="?→N
(B-A)÷N→D
0→S
A→X
For 1→K To N
S+f6D→S
X+D→X
Next
A→X
f6 →E
B→X
f6 →F
S+(F-E)D÷2→S
S
```

Simpson's Rule

The following program uses Simpson's Rule to approximate the value of a definite integral. The function you wish to integrate must be stored in the variable f_6 before you execute this program. You are also required to input the lower limit of integration (A), the upper limit of integration (B), and the number of subintervals (N). You must choose an even number for N. You can also use the calculator's built-in numerical integration program, which uses Simpson's Rule as its algorithm. See the Owner's Manual for more information on the built-in function.

```
"A="?→A
"B="?→B
"N="?→N
(B-A)÷N → D
0 → S
A → X
f6 → S
B → X
S+f6 → S
For 1→K To N÷2
A+(2K-1)D → X
S+4f6 → S
Next
For 1→K To (N÷2)-1
A+2KD → X
S+2f6 → S
Next
SD÷3
```

Integration by Endpoints or Midpoint

The following programs approximate the value of a definite integral. The function you wish to integrate must be stored in the Function Memory as f_6 before you execute this program. You are also required to input the lower limit of integration (A), the upper limit of integration (B), and the number of subintervals (N).

Left Endpoints:

```
"A="?→A
"B="?→B
"N="?→N
(B-A)÷N→D
0→S
A→X
For 1→K To N
S+f6D→S
X+D→X
Next
S
```

Right Endpoints:

```
"A="?→A
"B="?→B
"N="?→N
(B-A)÷N→D
0→S
A→X
X+D→X
For 1→K To N
S+f6D→S
X+D→X
Next
S
```

Midpoints:

```
"A="?→A
"B="?→B
"N="?→N
(B-A)÷N→D
0→S
A→X
X+.5D→X
For 1→K To N
S+f6D→S
```

```
X+D→X
Next
S
```

Euler's Method

The following program performs Euler's method to approximate the solution to the differential equation

$$\frac{dy}{dx} = f(x).$$

The function $f(x)$ must be stored as f_6 before you execute this program. You also must enter an initial point (both x and y coordinates), the increment between successive x values, and the value of x at which you wish to find an estimate for y.

```
"INITIAL X VALUE"?→X
"INITIAL Y VALUE"?→Y
"INCREMENT"?→H
"FINAL X VALUE"?→Z
Lbl 1
Y+H× f6 → Y
X+H → X
X≠Z ⇒ Goto 1
"FINAL X VALUE"
X ▲
"FINAL Y VALUE"
Y
```

Normal Probability

The following program computes the probability that a normal random variable with given mean and standard deviation lies between two given values. The probability that the random variable lies below a particular value may be computed by inputting -1E99 for the lower limit; the probability that the random variable lies above a particular value may be computed by inputting 1E99 for the upper limit.

```
"MEAN"?→M
"STD DEV"?→S
"LOWER LIMIT"?→A
(A-M)÷S→A
If A<-10
```

```
Then -10→A
IfEnd
If A>10
Then 10→A
IfEnd
"UPPER LIMIT"?→B
(B-M)÷S→B
If B<-10
Then -10→B
IfEnd
If B>10
Then 10→B
IfEnd
"PR(A<N(M,S)<B)="
Fix 4
∫(e(X^2÷-2),A,B)÷√(2×π)
```

Programs for the HP-38G

Introduction

The following section contains programs for the HP-38G calculator. These programs allow the calculator to do various financial calculations, linear programming, numerical integration, Euler's method for the numerical solution of differential equations, and to find probabilities associated with the normal distribution.

Entering the Programs Manually: All editing of programs is done within the Program catalog; to enter the Program catalog, press PROGRAM (the blue version of the 0 key). To enter a new program, press the menu key below NEW on the calculator screen; to edit an existing program, use the arrow keys to move down the list of programs until the one you wish to edit is shaded, then press the menu key below EDIT on the calculator screen. If you are creating a new program, you will be prompted for a program name. After entering a name and pressing OK twice, you will be ready to begin entering the program.

There are several places to go to find the characters you will need to enter the programs. Some arithmetic commands are available from the keyboard, as are the alphabet keys. Upper-case letters are entered by pressing the A...Z key (which is just above the blue key on the keypad) then the desired letter; lower-case letters are entered as upper-case, except you press the blue key before beginning the process. To lock the calculator in upper-case alphabet entry mode, you press the menu key below A...Z on the calculator screen; to lock the calculator in lower-case entry mode, press the blue key then the menu key below A...Z. You can easily tell when the calculator is in alphabet entry mode by the presence of an α at the very top of the screen. You can tell when it is locked in alphabet mode by the presence of a square in the A...Z menu key. The ▶ symbol you will need for these programs is the menu key below STO▶ on the calculator screen; keys to enter a space or to backspace are located at the menu keys below SPACE and BKSP on the calculator screen.

In addition to the keyboard symbols, you will need special characters like quotation marks and the summation sign. You will find these symbols by pressing CHARS (the blue version of the (key). Use the arrow keys to shade the symbol you want, then press ENTER; that symbol will be placed into the program at the point where the cursor was last. For commands like `INPUT`, `ROUND`, and `MSGBOX`, you may either enter them letter by letter or by pressing MATH. To find mathematical commands like ROUND, highlight the Real menu, then move over to the right hand side of the box and move down the list until you highlight ROUND. To find data input and output commands, press the menu key below CMDS on the calculator screen. Move down the left hand side of the box until you highlight Prompt, then move to the right hand side of the box and move down the list until you higlight the appropriate command. Pressing ENTER will return the command to the program at the cursor. The User's Guide contains the menu locations of all available commands.

When reading the programs in the following section, you will notice a seemingly arbitrary amount of horizontal space in the MSGBOX lines of the programs. This space is not arbitrary, but is designed to make the output of the programs to be as readable as possible. You may estimate the number of spaces necessary to fill these gaps, and enter them into the programs. To exit the Program catalog, you may press HOME to return to the home screen.

Running the Programs: You may run the programs from either the home screen or from the Program catalog. To run a program from the catalog, highlight its name using the arrow keys then press the menu key below RUN on the calculator screen. To run a program from the home screen, type RUN followed by the program name. What you see will depend on from where you have run the program. If you run the program from the home screen, the final output of the program will be returned to the screen and will stay there even after the program has terminated and you have pressed OK to return to the home screen. If you run the program from the Program catalog, the output will disappear after you press OK, and you will be returned to the Program catalog screen.

Future Value of an Annuity

The following program computes the future value of an annuity given the payment amount, the rate (remember that percents should be converted to decimals), and the number of payments. Payments are made at the end of each period.

```
INPUT R;
"FUTURE VALUE OF AN ANNUITY";
"PAYMENT";
"ENTER PAYMENT";
Ø:
INPUT I;
"FUTURE VALUE OF AN ANNUITY";
"RATE";
"ENTER RATE (IN DECIMAL FORM)";
Ø:
INPUT N;
"FUTURE VALUE OF AN ANNUITY";
"# PYMNTS";
```

```
"ENTER NUMBER OF PAYMENTS";
Ø:
ROUND(R*((1+I)^N-1)/I,2)▶V:
MSGBOX "Future Value:      $" V:
FREEZE:
```

Present Value of an Annuity

The following program computes the present value of an annuity given the payment amount, the rate (remember that percents should be converted to decimals), and the number of payments. Payments are made at the end of each period.

```
INPUT R;
"PRESENT VALUE OF AN ANNUITY";
"PAYMENT";
"ENTER PAYMENT";
Ø:
INPUT I;
"PRESENT VALUE OF AN ANNUITY";
"RATE";
"ENTER RATE (IN DECIMAL FORM)";
Ø:
INPUT N;
"PRESENT VALUE OF AN ANNUITY";
"# PYMNTS";
"ENTER NUMBER OF PAYMENTS";
Ø:
ROUND(R*(1-(1+I)^(-N))/I,2)▶V:
MSGBOX "Present Value:    $" V:
FREEZE:
```

Amortization Table

The following program asks for the following information about a loan: the starting balance, the rate per period (percents should be converted to decimals), amount of each regular payment, and the number of payments. The program displays an amortization table for this loan in the form of a matrix through which you can scroll by using the arrow keys. In this table the columns are respectively:

Payment Number	Amount of Payment	Interest for the Period	Portion to Principal	New Balance

After scrolling through the table, pressing the OK menu key displays the total payments made; pressing OK again displays the total interest paid. The program stores the table as matrix MØ for future viewing (until the program is run again and a new table takes its place). When the regular payment is too small, the final payment may be quite large. When the regular payment is too large, the final payment may occur before N is reached, where N is the number of payments entered at the beginning of the program. In this case the matrix will have fewer than N rows. You should also note that the values of N which you may use are limited by the amount of available memory in the calculator; if you exceed this number you will receive the error message `Insufficient Memory; Edit Program?`.

```
INPUT B;
"AMORTIZATION";
"BALANCE";
"ENTER STARTING BALANCE";
Ø:
INPUT I;
"AMORTIZATION";
"RATE";
"ENTER RATE PER PERIOD";
Ø:
INPUT P;
"AMORTIZATION";
"PYMNT";
"ENTER AMOUNT OF PAYMENT";
Ø:
INPUT N;
"AMORTIZATION";
"# PYMNTS";
"ENTER NUMBER OF PAYMENTS";
Ø:
MAKEMAT(Ø,N,5)▶MØ:
1▶K:
Ø▶W:
WHILE K<N AND B>P REPEAT
K▶MØ(K,1):
P▶MØ(K,2):
ROUND(I*B,2)▶MØ(K,3):
MØ(K,3)+W▶W:
P-MØ(K,3)▶MØ(K,4):
```

```
B-MØ(K,4)►MØ(K,5):
MØ(K,5)►B:
K+1►K:
END:
K►MØ(K,1):
B+ROUND(I*B,2)►MØ(K,2):
ROUND(I*B,2)►MØ(K,3):
MØ(K,2)-MØ(K,3)►MØ(K,4):
B-MØ(K,4)►MØ(K,5):
REDIM MØ;{K,5}:
EDITMAT MØ:
(K-1)*P+MØ(K,2)►U:
W+MØ(K,3)►V:
MSGBOX "Total Payments:  $"U:
MSGBOX "Total Interest:  $"V:
FREEZE:
```

Linear Programming – Maximization

The following program performs the simplex method on a tableau (matrix) which has been previously input as MØ. To run the program you should store your initial tableau as MØ. The final matrix is stored as M1 and is displayed at the end of the program. This final matrix may be investigated by scrolling using the arrow keys; when you are finished scrolling, pressing the OK menu key returns you to the Program catalog.

```
SIZE(MØ)►LØ:
LØ(1)▷R:
LØ(2)►C:
MØ►M1:
1►P:
WHILE P≠Ø REPEAT
Ø►P:
TRN(M1)►M2:
FOR K=1 TO C-1 STEP 1;
IF M1(R,K)<Ø THEN
2►P:
END:
END:
FOR K=1 TO R-1 STEP 1;
IF M1(K,C)<Ø THEN
```

```
1►P:
END:
END:
IF P==1 THEN
MAKELIST(M1(X,C),X,1,R,1)►L6:
Ø►L6(R):
SORT(L6)►L1:
L1(1)►M:
FOR K=1 TO R-1 STEP 1;
IF L6(K)==M THEN
K►I:
END:
END:
MAKELIST(M2(X,I),X,1,C,1)►L6:
Ø►L6(C):
SORT(L6)►L1:
L1(1)►M:
IF M<Ø THEN
FOR K=1 TO C-1 STEP 1;
IF L6(K)==M THEN
K►J:
END:
END:
ELSE Ø►P:
END:
END:
IF P==2 THEN
MAKELIST(M2(X,R),X,1,C,1)►L6:
Ø►L6(C):
SORT(L6)►L1:
L1(1)►M:
FOR K=1 TO C-1 STEP 1;
IF L6(K)==M THEN
K►J:
END:
END:
MAKELIST(M1(X,J),X,1,R,1)►L6:
Ø►L6(R):
FOR K=1 TO R-1 STEP 1;
IF L6(K)>Ø THEN
L6(K)/M1(K,C)►L6(K):
```

```
ELSE Ø▶L6(K):
END:
END:
SORT(L6)▶L1:
L1(R)▶M:
IF M>Ø THEN
FOR K=1 TO R-1 STEP 1;
IF L6(K)==M THEN
K▶I:
END:
END:
ELSE Ø▶P:
END:
END:
IF P≠Ø THEN
SCALE M1;1/M1(I,J);I:
FOR K=1 TO R STEP 1;
IF K≠I THEN
SCALEADD
M1; -M1(K,J);I;K:
END:
END:
END:
END:
EDITMAT M1:
```

Trapezoidal Rule

The following program uses the Trapezoidal Rule to approximate the value of a definite integral. The function you wish to integrate must be stored in the variable `F1(X)` before you execute this program. To do this, press the LIB key, note that `Function` is highlighted, and press ENTER. You may now enter the function `F1(X)`. You are also required to input the lower limit of integration (`A`), the upper limit of integration (`B`), and the number of subintervals (`N`).

```
INPUT A;
"TRAPEZOIDAL RULE";
"A";
"ENTER LOWER LIMIT";
Ø:
INPUT B;
```

```
"TRAPEZOIDAL RULE";
"B";
"ENTER UPPER LIMIT";
Ø:
INPUT N;
"TRAPEZOIDAL RULE";
"N";
"ENTER NUMBER OF INTERVALS";
Ø:
(B-A)/N▶D:
F1(A)+F1(B)▶S:
S+2*∑(I=1,N-1,F1(A+I*D))▶S:
MSGBOX "Trapezoidal Rule Sum:           "S*D/2:
FREEZE:
```

Simpson's Rule

The following program uses Simpson's Rule to approximate the value of a definite integral. The function you wish to integrate must be stored in the variable `F1(X)` before you execute this program. To do this, press the `LIB` key, note that `Function` is highlighted, and press `ENTER`. You may now enter the function `F1(X)`. You are also required to input the lower limit of integration (`A`), the upper limit of integration (`B`), and the number of subintervals (`N`). You must choose an even number for `N`.

```
INPUT A;
"SIMPSON'S RULE";
"A";
"ENTER LOWER LIMIT";
Ø:
INPUT B;
"SIMPSON'S RULE";
"B";
"ENTER UPPER LIMIT";
Ø:
INPUT N;
"SIMPSON'S RULE";
"N";
"ENTER (EVEN) NUMBER OF INTERVALS";
Ø:
(B-A)/N▶D:
```

```
F1(A)+F1(B)▶S:
S+4*∑(I=1,N/2,F1(A+(2*I-1)*D)▶S:
S+2*∑(I=1,(N/2)-1,F1(A+2*I*D)▶S:
MSGBOX "Simpson's Rule Sum:                    "S*D/3:
FREEZE:
```

Integration by Endpoints or Midpoint

The following programs approximate the value of a definite integral. The function you wish to integrate must be stored in the variable `F1(X)` before you execute this program. To do this, press the `LIB` key, note that `Function` is highlighted, and press `ENTER`. You may now enter the function `F1(X)`. You are also required to input the lower limit of integration (`A`), the upper limit of integration (`B`), and the number of subintervals (`N`).

Left Endpoints:

```
INPUT A;
"LEFT ENDPOINT SUM";
"A";
"ENTER LOWER LIMIT";
Ø:
INPUT B;
"LEFT ENDPOINT SUM";
"B";
"ENTER UPPER LIMIT";
Ø:
INPUT N;
"LEFT ENDPOINT SUM";
"N";
"ENTER NUMBER OF INTERVALS";
Ø:
(B-A)/N▶D:
∑(I=Ø,N-1,F1(A+I*D)▶S:
MSGBOX "Left Endpoint Sum:                    "S*D:
FREEZE:
```

Right Endpoints:

```
INPUT A;
"RIGHT ENDPOINT SUM";
"A";
"ENTER LOWER LIMIT";
Ø:
INPUT B;
"RIGHT ENDPOINT SUM";
"B";
"ENTER UPPER LIMIT";
Ø:
INPUT N;
"RIGHT ENDPOINT SUM";
"N";
"ENTER NUMBER OF INTERVALS";
Ø:
(B-A)/N▶D:
∑(I=1,N,F1(A+I*D)▶S:
MSGBOX "Right Endpoint Sum:    "S*D:
FREEZE:
```

Midpoints:

```
INPUT A;
"MIDPOINT SUM";
"A";
"ENTER LOWER LIMIT";
Ø:
INPUT B;
"MIDPOINT SUM";
"B";
"ENTER UPPER LIMIT";
Ø:
INPUT N;
"MIDPOINT SUM";
"N";
"ENTER NUMBER OF INTERVALS";
Ø:
(B-A)/N▶D:
∑(I=1,N,F1(A+(I-.5)*D)▶S:
MSGBOX "Midpoint Sum:    "S*D:
FREEZE:
```

Euler's Method

The following program performs Euler's method to approximate the solution to the differential equation

$$\frac{dy}{dx} = f(x).$$

The function $f(x)$ must be stored as `F1(X)` before you execute this program. To do this, press the `LIB` key, note that `Function` is highlighted, and press `ENTER`. You may now enter the function `F1(X)`. You also must enter an initial point (both x and y coordinates), the increment between successive x values, and the value of x at which you wish to find an estimate for y.

```
INPUT X;
"EULER'S METHOD";
"XØ";
"ENTER INITIAL X VALUE";
Ø:
INPUT Y;
"EULER'S METHOD";
"YØ";
"ENTER INITIAL Y VALUE";
Ø:
INPUT H;
"EULER'S METHOD";
"INCR";
"ENTER INCREMENT";
Ø:
INPUT Z;
"EULER'S METHOD";
"FINAL X";
"ENTER FINAL X VALUE";
Ø:
WHILE X≠Z REPEAT
Y+H*F1(X)▶Y:
X+H▶X:
END:
DISP 2; "FINAL X VALUE:":
DISP 3; "  "X:
DISP 4; "FINAL Y VALUE:":
```

```
DISP 5; " "Y:
FREEZE:
```

Normal Probability

The following program computes the probability that a normal random variable with given mean and standard deviation lies between two given values. The probability that the random variable lies below a particular value may be computed by inputting -1E99 for the lower limit; the probability that the random variable lies above a particular value may be computed by inputting 1E99 for the upper limit.

```
INPUT M;
"NORMAL PROBABILITY";
"MEAN";
"ENTER MEAN";
Ø:
INPUT S;
"NORMAL PROBABILITY";
"STD DEV";
"ENTER STANDARD DEVIATION";
Ø:
INPUT A;
"NORMAL PROBABILITY";
"A";
"ENTER LOWER LIMIT";
Ø:
INPUT B;
"NORMAL PROBABILITY";
"B";
"ENTER UPPER LIMIT";
Ø:
UTPN(M,S^2,A)-UTPN(M,S^2,B)▶V:
MSGBOX "Pr(A<N(M,S)<B)=" V:
FREEZE:
```

Part IV

General Instructions for Excel

IV-2 Spreadsheets

In this part, we give an overview of the general concepts and tools that involve spreadsheets. The spreadsheet software we use is Microsoft's Excel, but the principles should hold for any current software package.

Spreadsheets

Introduction

One could argue that the advent of spreadsheet and word-processing software ushered in the PC as a primary tool in every workplace. In today's world, wherever there is a table of data, it is stored in a spreadsheet. Some of the things one can do with data via spreadsheets are:

- Data Arrangement
- Calculation
- Visualization
- Database Management
- Statistical Analysis
- Predictions
- Optimization

Spreadsheet packages also come with powerful built-in programming languages, making their versatility almost limitless. The best part about spreadsheets is that the initial learning curve is very short! We use Microsoft Excel 2000 for all demonstrations in this manual, but due to the standardization in today's spreadsheets, one should be able to apply this material to almost any spreadsheet package. Throughout this manual, spreadsheet and PC terminology will be used. In Figure 1, the typical window one sees is displayed with many of the names we use in this manual.

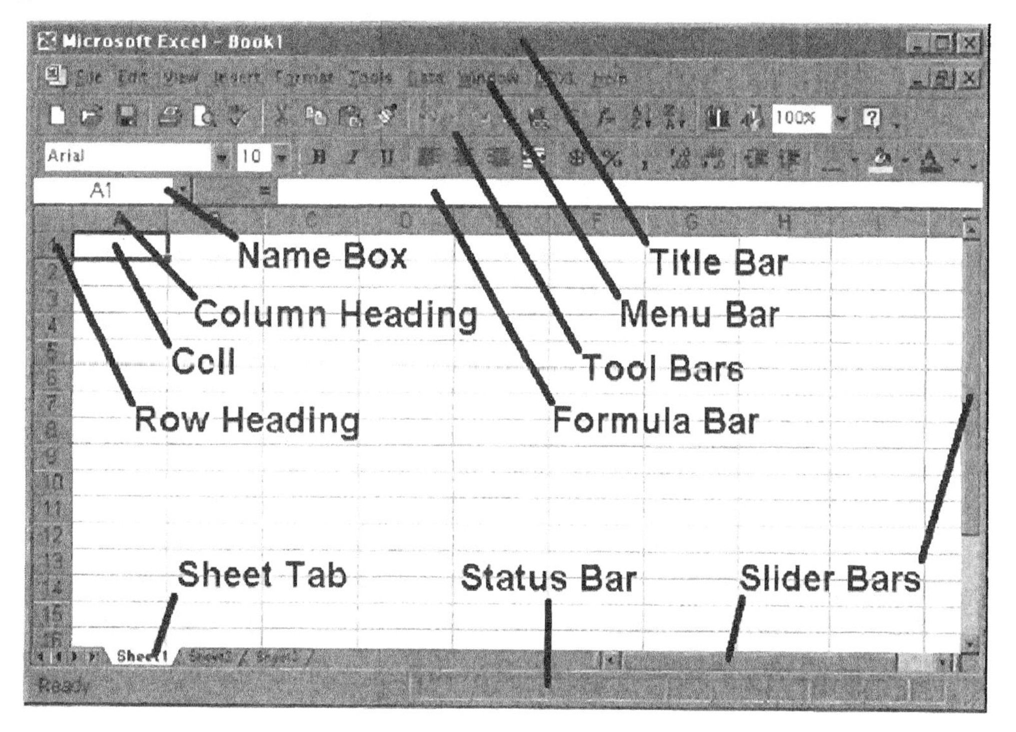

Figure 1: These names should help keep you straight.

Files

Collections of sheets are stored in files called **workbooks**. When you open Excel, it starts with a brand new workbook called "Book1.xls". Figure 1 is very close to what you will see. Notice the title of the workbook in the **title bar** at the top of the window. If you want to open a previously saved workbook, you can do so through **File → Open** on the **menu bar** or open it directly by finding the file and double-clicking it.

Each workbook consists of sheets, accessed via the **sheet tabs** at the bottom of the screen (Figure 1). Each sheet can be either a **chart** or a **worksheet**. Charts are any kind of visualization of data, such as scatter plots or bar graphs. Worksheets are the "spreadsheets", the place where we (and the program) do all of the work.

Cells

Everything begins with the **cell**. If you can understand exactly what a cell is and what properties it has, the rest should be relatively easy. *It is crucial that you understand this section!*

A cell is one of the many rectangles you see on a worksheet. One or more cells are **selected** when they are surrounded by a bold outline with a **fill handle** in the lower right hand corner of the outline (we will discuss the purpose of the fill handle later). You can select a single cell by clicking on it or select a rectangular set of them by click-dragging (Figure 2)

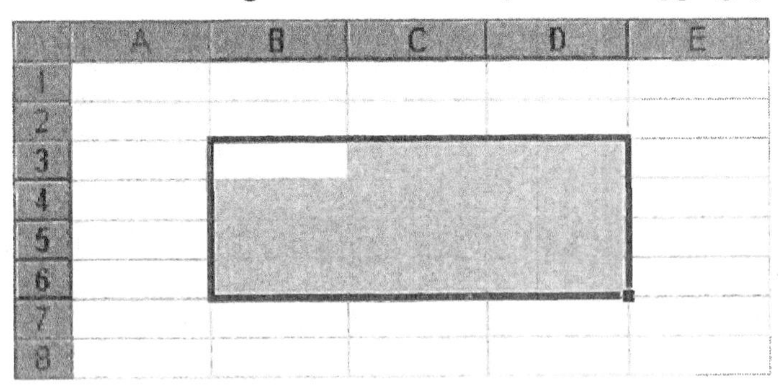

Figure 2: The range `B3:D6` is selected.

The cell's primary function is to hold and display "stuff". *The cell has four major attributes, which you must learn:*

- Address
- Format
- Content
- Value

Cell Address

The **address** of a cell is indicated by the row and column in which it sits. (If you have ever played Battleship®, you know exactly how this works!) The **column heading** is always one or two letters (ranging from `A` to `IV`) and the **row heading** is always a number (from `1` to `65536`). A cell's address consists of the column letter(s) and the row number. For example, if a cell sits in column B and row 3, then its address is B3 (Figure 3). Note: The address of the selected cell is given in the **name box**.

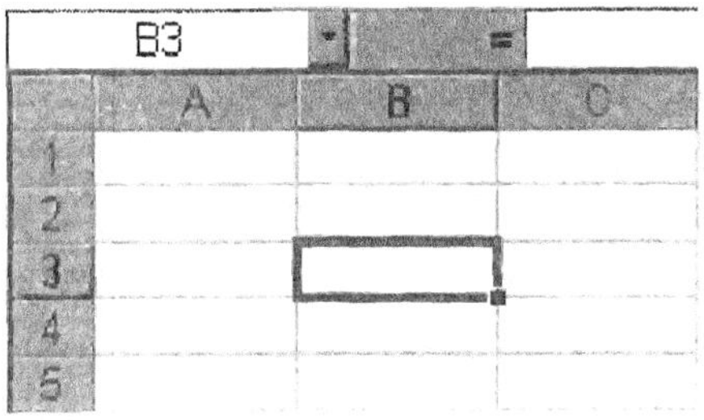

Figure 3: The address of the selected cell is B3

When a rectangular **range** of cells is selected, the address of the range is given by the address of the upper left cell and the lower right cell, separated by a colon. In Figure 2 the upper left cell of the selected range is B3 and the lower right cell is D6, thus the range is denoted by B3:D6. Note: When a range of cells forms a single column or row, the two end cells are used to denote the range.

Cell Format

A cell can contain many things, like numbers and text, but how the cell displays those things can vary greatly. Suppose you type into every cell of a range (say B1:B5) the number 0.75. Without any formatting, the cells all display 0.75. By right-clicking a cell, you bring up a plethora of options, including **Format Cells...** (like most everything else, this can also be obtained through the menu bar). The first tab you get in the **Format Cells** pop-up window is **Number** (Figure 4). Here is where you can format the cell to display its value as anything from dates to currency. You can even dictate how many decimal places the cell will display (which can cause Excel to round the displayed value but *not* the actual value). In Figure 5 you can see some of the possibilities, where all the cells contain the same value 0.75.

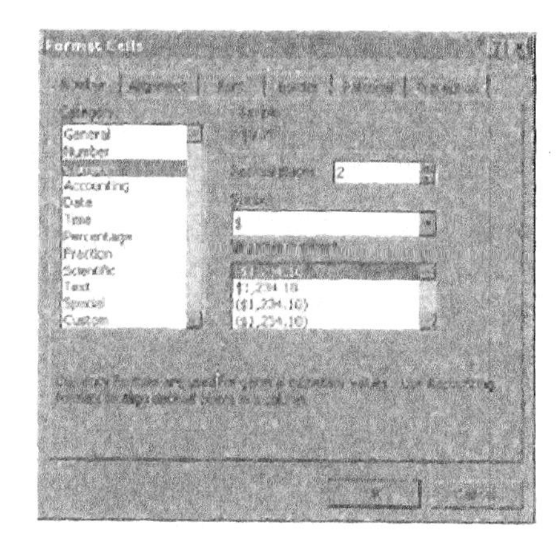

Figure 4: Format Cells pop-up

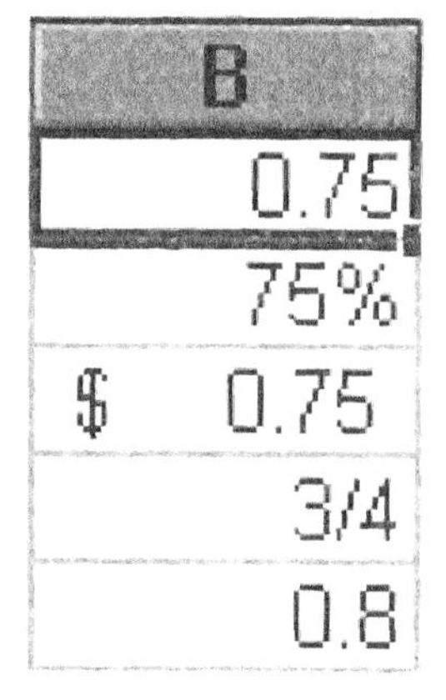

Figure 5: Assorted formatting of 0.75.

Cell Content

The **content** of a cell is whatever you type into it, which is usually a number, text, or a formula. The content of the cell is not necessarily what the cell displays! In other words, you may type a formula into a cell and press Enter, then what the cell *displays* is the value of the formula, while the *content* of the cell is the formula. Whenever you select a cell, the content of the cell is shown in the **formula bar**. For example suppose you select a cell, type the formula =3/4, and press Enter. If you formatted the cell as a percentage, then the cell will display the value 75%. Now reselect cell, look in the formula bar and you should see the formula =3/4 (Figure 6).

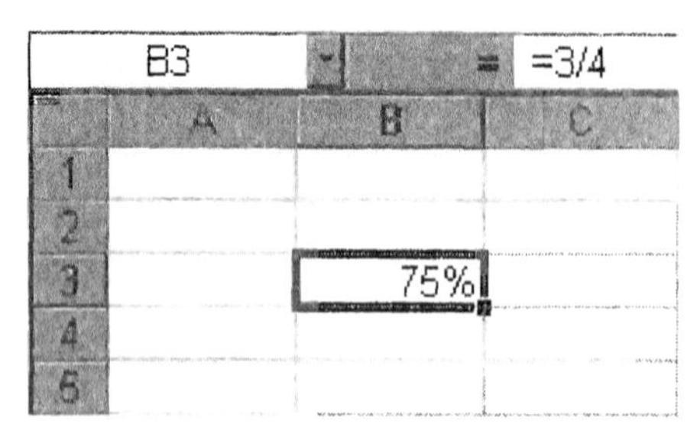

Figure 6: Cell B3 displays 75%, has a value of 0.75, and contains the formula =3/4.

This is what makes spreadsheets so powerful. You will be able to enter formulas into cells that use values from other cells. In turn, these other cells may contain formulas themselves!

There is something very important to remember when using longer, more complicated formulas in Excel. *Excel does not correctly obey the standard order of operations!* Recall from basic algebra that $-3^2 = -9$ and $(-3)^2 = 9$, because an exponent applies *only* to whatever is to its immediate left. Unfortunately, Excel is programmed to apply negation (the negative sign) *before* applying exponents, whether parentheses are present or not. So the formula =-A1^2 will return the value of -A1 squared, not the negative of A1 squared. Therefore, when using Excel, you should adopt the convention of enclosing the base *and* the exponent in parentheses. Table 1 shows the result of three different formulas. If you wish to calculate -3^2, then you should use the formula shown in the last column of the table.

A2	=A2^2	=-A2^2	=-(A2^2)
3	9	9	-9

Table 1.

Cell Value

Every cell has a numeric **value**, even an empty one (which has a default value of 0). If you type a number into a cell, then that will also be the cell's value. If you type a numeric formula into a cell, then that cell's value will be the outcome of the formula. When the cell contains text or has a formula that returns text, then the value of that cell will default to 0. Remember, the value of a cell, what the cell displays, and the cell's content are three different things. As seen in the example illustrated in Figure 6, the value of the cell is a 0.75, while the cell's content is the formula =3/4 and the cell's display is 75%.

Formulas

As seen earlier, you can type formulas into a cell. All formulas start with an "=". A formula can have numbers, algebraic operations, **functions**, and addresses for cells and ranges. Notice that we do not have variables! Any time we want to use a value from another cell in a formula, we can use the cell's address instead. Let us consider an example.

Suppose you wanted to store in cell `D2` the average of three numbers that you have stored in cells `B1`, `B2` and `B3`. This can be done with a formula. In cell `D2`, type `=(B1+B2+B3)/3` and press `Enter`. The cell `D2` displays the average. If you reselect cell `D2`, the formula bar will display the formula that the cell contains, while the cell displays its value (Figure 7).

Figure 7: Average of cells via a simple formula.

For many common calculations (such as averaging) there are built-in functions. To view a list of these functions, click on a cell and type "=". The name box changes and becomes a drop-down list of functions you have used recently, but right now the most interesting choice is **More Functions...** (Figure 8). Through this choice, you can explore and use a whole host of functions; it is certainly worth your while to check this out! Another way of obtaining a function is through the menu bar **Insert → Function**. You can also type in functions directly if you know the correct spelling and appropriate arguments. In Figure 9, the function `=AVERAGE(B1:B3)` gives the same result as seen in Figure 7.

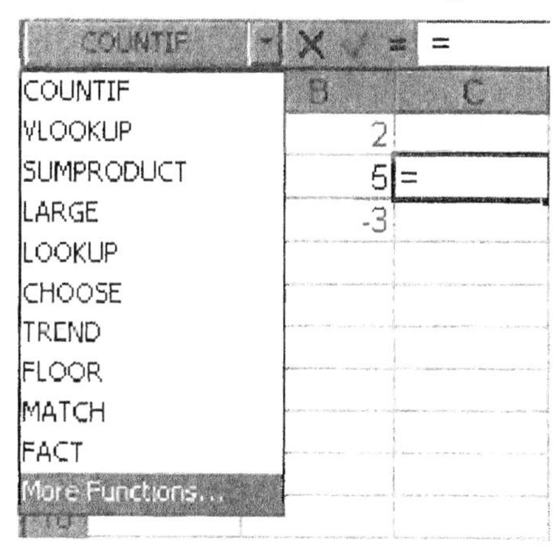

Figure 8: Average of cells via a simple formula.

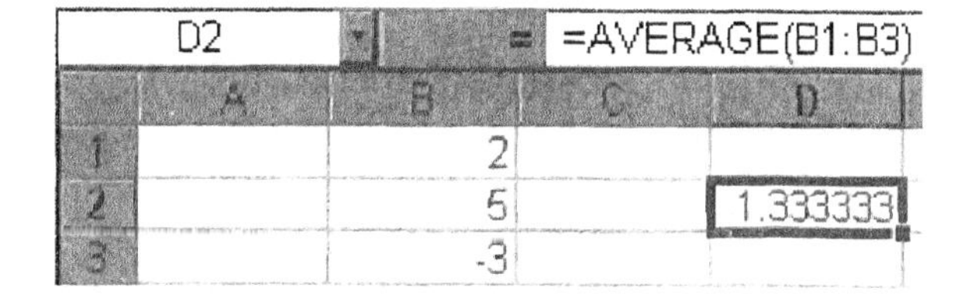

Figure 9: Average of cells via a simple formula.

Copy, Paste, and Fill

Anyone who has used a PC is probably quite familiar with cut, copy, and paste. When copying cells, if the cell's content is anything other than a formula, the contents are identically transcribed. However, when the content of a copied cell is a function, then the addresses within the function are adjusted by the relative distance from the copied cell to the pasted cell.

As an example, consider Figure 10 where you have a table of values of which you want the average of each column. Type in the appropriate formula for averaging the first column, copy that cell and paste into the cell under the next column (Figure 11). Notice how the address range A1:A3 is shifted by exactly one column value to B1:B3. This is precisely the relative change from the copied cell (A4) to the pasted cell (B4). If you pasted from the cell A4 to the cell C9, the relative change would be two to the left and five down, thus the pasted formula

A4 =AVERAGE(A1:A3)

	A	B	C	D
1	3	2	5	8
2	7	5	-9	8
3	2	-3	2	3
4	4			

Figure 10: The first average function is typed in.

B4 =AVERAGE(B1:B3)

	A	B	C	D
1	3	2	5	8
2	7	5	-9	8
3	2	-3	2	3
4	4	1.333333		

Figure 11: The next average is copied from the first.

B4 =AVERAGE(B1:B3)

	A	B	C	D
1	3	2	5	8
2	7	5	-9	8
3	2	-3	2	3
4	4	1.333333		

Figure 12: The fill handle is grabbed and dragged.

B4 =AVERAGE(B1:B3)

	A	B	C	D
1	3	2	5	8
2	7	5	-9	8
3	2	-3	2	3
4	4	1.333333	-0.66667	6.333333

Figure 13: All of the averages are correctly filled in.

would be =AVERAGE(C6:C8).

Coming back to our example, there is a much quicker way of copying one cell to multiple cells. The technique is called **filling**, and with most things, there are several ways of accomplishing it. As noted earlier, every highlighted cell has a fill handle in the lower right-hand corner. If you hold-click this handle (grab) and drag in any direction, an automatic copy/paste is performed on all covered cells. In Figure 12 the handle on cell B4 is shown grabbed and dragged across cells C4 and D4. When the mouse is released, the average function is pasted into the both C4 and D4 with the appropriate adjustment to the address ranges within the function (Figure 13).

There may be times when you will copy/paste a cell and you want one or more parts of the address ranges in the function's arguments to *not* change. The trick is to use a $ in front of every part of the addresses you want to freeze. For instance, suppose you are filling from a cell with the formula =A1+B3-C7. If you do not want the column of the address A1 to change, you would instead use the formula =$A1+B3-C7. If you do not want the row address of B3 to change, you would use the formula =A1+B$3-C7. If you do not want the address C7 to change at all, you would use the formula =A1+B3-$C$7. When an address is completely frozen, e.g. $C$7, this is called an **absolute reference**. When no part of the cell reference is frozen, e.g. C7, this is called a **relative reference**. When part of a cell reference is frozen, e.g. $C7 or C$7, this is called a **mixed reference**.

Charts

A picture is worth a thousand words. Spreadsheets are loaded with graphical tools for displaying sets of data. The best part is how easy it is to create a **chart** of data. Consider the data in Figure 14 of homework and test grades over the first four chapters of a class. Select the data to be visualized (range `B1:C5`) and engage the **chart wizard** via the tool bar or the menu under **Insert**. As seen in Figure 15, the first screen of the chart wizard gives many choices of possible charts, including bar charts, pie charts, and scatter plots. The default choice is the standard grouped bar graph.

	A	B	C
1	Chapter	HW	Test
2	1	75	73
3	2	82	78
4	3	78	80
5	4	91	88

Figure 14: These are class grades to be visualized.

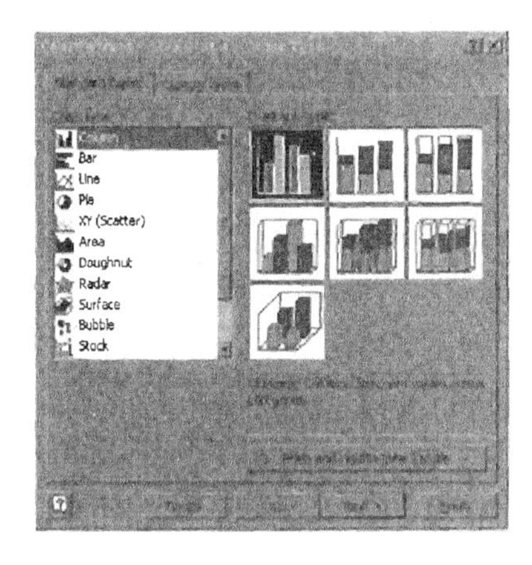

Figure 15: There are a host of choices in charts.

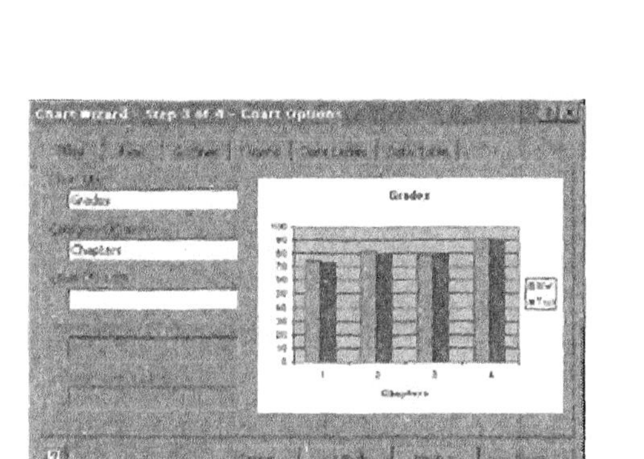

Figure 16: The chart wizards gives lots of control over the chart.

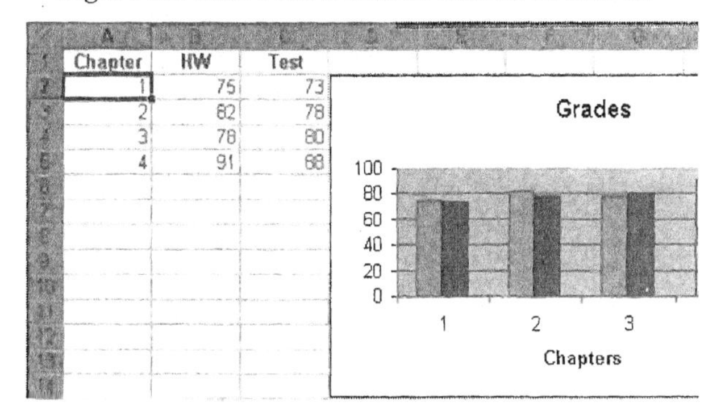

Figure 17: The chart can be added to the data spreadsheet or placed on a separate sheet.

As you move through the wizard, you are given a vast number of choices to fine tune and annotate your graph (Figure 16). The last choice you need to make is whether or not the chart will be displayed on the data spreadsheet or given a sheet of its own (Figure 17).

After the chart is created, changes can be made by simply right-clicking the part of the chart in which you are interested. For example, if you want to change the format of the title, right-click it and select **Format Chart Title** from the resulting menu. You can make major changes to the whole chart by right-clicking in the white area of the chart, outside of any parts.

The resulting menu allows you to change everything from chart type to gridlines. Later in this manual you will see how by right-clicking the data graphics, one can even include trendlines!

Conclusion.

Once you become comfortable with the spreadsheet environment, you should experiment with all the available charts, formulas, and built-in abilities. The possibilities for effectively and efficiently using Excel, or any other spreadsheet program, are practically endless. If you are using Microsoft Word or Microsoft PowerPoint, all tables and charts from Excel can be copied and pasted into a Word document or PowerPoint presentation so that you can create attractive papers and presentations. You can even import data from other programs and databases, as well as from the World Wide Web. Even though this manual only demonstrates how to use Excel for problems that arise in finite mathematics and in applied calculus, Excel also has applicability to other courses, including statistics, accounting, economics, biology, chemistry, and physics, just to name a few.

You may also find spreadsheets useful for keeping track of your college credits, personal finances, even as an address book for e-mail and mailing addresses and phone numbers. You are limited only by your imagination and patience.

Part V

Detailed Instructions for Excel

Part V contains detailed steps and instructions for using Microsoft's Excel to complete many examples from *Finite Mathematics*, ninth edition, *Calculus with Applications*, ninth edition, *Calculus with Applications: Brief Version*, ninth edition, and *Finite Mathematics and Calculus with Applications*, eighth edition. Most of the instructions provided can be applied to other current spreadsheet programs.

Detailed Instructions for Finite Mathematics

This section of Part V contains detailed instructions for using Microsoft Excel for *Finite Mathematics*, ninth edition, and related chapters in *Finite Mathematics and Calculus with Applications*, eighth edition. Chapter 1 is common to these two texts, as well as *Calculus with Applications*, ninth edition, and *Calculus with Applications: Brief Version*, ninth edition, This section is organized by chapters in *Finite Mathematics*; since not all chapters require detailed explanations of spreadsheet use, some chapters are not mentioned here.

In this manual, section titles from the textbooks are indicated in italics. References are made to specific examples and exercises from the corresponding sections of each chapter, so you should have your textbook nearby as you read through these instructions.

Chapter 1 Linear Functions

LOCATION IN THE OTHER TEXTS:

Calculus with Applications: Chapter 1
Calculus with Applications, Brief Version: Chapter 1
Finite Mathematics and Calculus with Applications: Chapter 1

Linear Functions and Applications.

Using *Goal Seek* to Solve Equations.

If an equation contains only one variable, then the **Goal Seek** function can solve it. Begin by creating a table containing the relevant variable and formulas. For instance, to solve Example 6(a) of your text with Excel, beginning in cell `A1`, set up appropriate column headings to represent the units (x), revenue, cost, and profit. Cell `A2` will serve as the variable cell, holding the place of x in our formulas. You need not enter anything in this cell. In the other cells in row 2, type the following:

Cell	Contents
`B2`	`=24*A2`
`C2`	`=20*A2+100`
`D2`	`=B2-C2`

The table should appear similar to Table 1.

Units (x)	R=24x	C=20x+100	Profit=R-C
0	0	100	-100

Table 1.

Access the **Goal Seek** function in the **Tools** menu. For the cell to "set", type `D2`; for the value to set this cell to, type `0`; the cell to be "changed" is the cell acting as our variable, which is cell `A2`. The **Goal Seek** pop-up should look like Figure 1. (Note: **Goal Seek** will automatically change all cell references entered into absolute references; see Part I of this manual for more information on absolute references.)

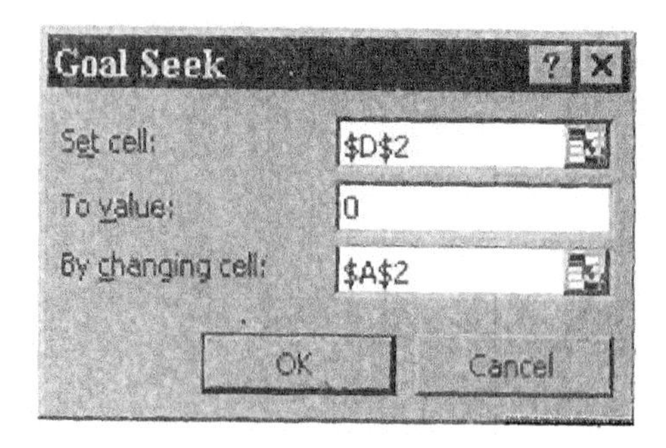

Figure 1.

Click **OK** and cell `A2` (the number of units) will be adjusted until cell `D2` (the profit) is 0. We see in Figure 2 that 25 units must be sold for this company to break-even.

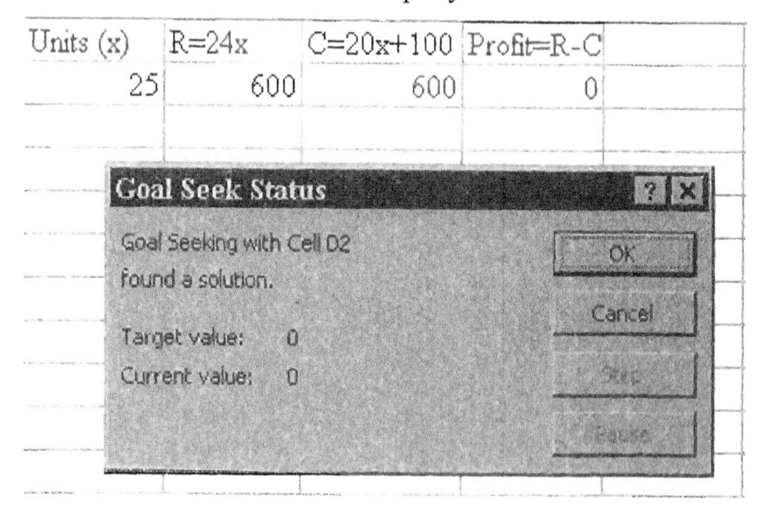

Figure 2.

The Least Squares Line

Finding the Least Squares Regression Line.

To complete the example in this section, begin by entering the lists of data points into columns `A` and `B`. We can now use the functions **SLOPE**, **INTERCEPT**, and **CORREL** to find m, b, and r, respectively. Type the following into the indicated cells:

Cell	Contents
`E1`	`=SLOPE(B2:B11,A2:A11)`
`E3`	`=INTERCEPT(B2:B11,A2:A11)`
`E5`	`=CORREL(B2:B11,A2:A11)`

When using the **SLOPE** and **INTERCEPT** functions, the list containing the y-values must be first, as shown above. The final result should resemble Table 2 on the next page.

Year	Death Rate		*Slope=*	-0.559697
10	84.4		*y-intercept=*	90.3333333
20	71.2		*r =*	-0.9629006
30	80.5			
40	73.4			
50	60.3			
60	52.1			
70	56.2			
80	46.5			
90	36.9			
100	34.0			

Table 2.

Displaying Data and The Regression Line on the Same Graph.

Begin by entering the lists of data points and creating an **XY (Scatter)** chart in the spreadsheet. (See Part I of this manual for additional assistance with creating charts.) Right-click on any data point in the chart and select **Add Trendline**. Select the first type, **Linear**, then click the **Options** tab. Check the boxes in front of "Display equation on chart" and "Display R-squared value on chart". Click **OK** and the least squares line and the resulting value of r^2 are displayed. You may need to click and drag the text box containing the equation and r^2 value to move it to a more convenient position within the chart. The final result should resemble Figure 3.

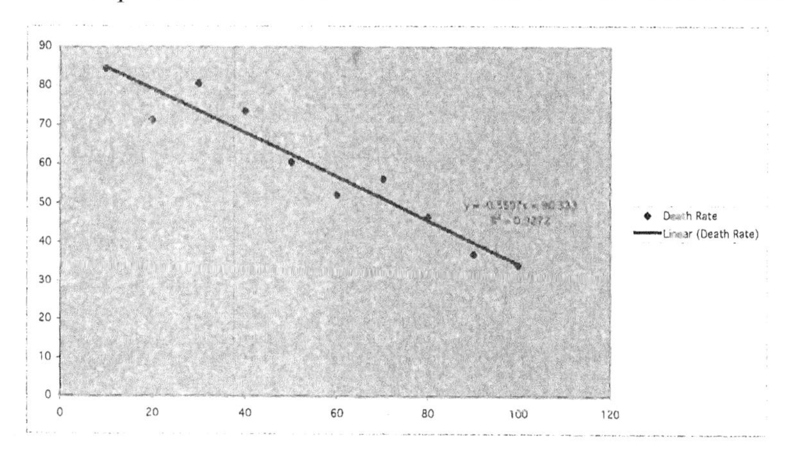

Figure 3.

It is important to remember that this method gives you r^2, which is the square of the correlation coefficient. To find the value of r, you must take the square root of this value, or use the CORREL function as described previously.

Chapter 2 Systems of Linear Equations and Matrices

LOCATION IN THE OTHER TEXTS:

Finite Mathematics and Calculus with Applications: Chapter 2

Solution of Linear Systems by the Gauss-Jordan Method.

Row Operations.

As indicated in your text, a clever use of the **Copy** and **Paste Special** commands can minimize the arithmetic required by the Gauss-Jordan Method for solving systems of linear equations. Begin by setting up the matrix, with appropriate row and column headings, in the cell range `A1:E4`, as shown in Table 1.

	x	y	z	Constant
R1	1	-1	5	-6
R2	3	3	-1	10
R3	1	3	2	5

Table 1.

Highlight the entire matrix, including the row and column headings, select **Copy** from the **Edit** menu, and click on cell `A6`. Select **Paste** from the **Edit** menu to paste the entire matrix. Since the first row operation we are asked to perform is $-3R_1 + R_2 \to R_2$, highlight the entire second row of the copied matrix and type in the following:

Cell	Contents
`B8`	`=-3*B2:E2+B3:E3`

Do not press **Enter**! Instead, press **Ctrl-Shift-Enter** (all three keys simultaneously). This will cause each of the cells in our copied R_2 to be updated with the values resulting from the row operation. To complete the next row operation, $-R_1 + R_3 \to R_3$, highlight the entire third row of the copied matrix and type in the following, pressing **Ctrl-Shift-Enter** when finished:

Cell	Contents
`B9`	`=-B2:E2+B4:E4`

The new matrix, with the updated rows, should look like Table 2.

	x	y	z	Constant
R1	1	-1	5	-6
R2	0	6	-16	28
R3	0	4	-3	11

Table 2.

Now, highlight the entire new matrix, and select **Copy** from the **Edit** menu. Click on cell `A11`, and this time, select **Paste Special** from the edit menu. In the window that pops up, select **Values** and click **OK**. This will copy only the values, not the formulas, from the previous matrix operations. To perform the next row operation from this example in your text, highlight the first row of the third matrix, type `=B8:E8+6*B7:E7`, and press **Ctrl-Shift-Enter** to update the new row. Notice that we are now creating our formulas from the second matrix, the one that was updated. To finish with the third matrix, select the third row, type `=2*B8:E8-3*B9:E9` and press **Ctrl-Shift-Enter**. This matrix should now look like Table 3.

	x	y	z	Constant
R1	6	0	14	-8
R2	0	6	-16	28
R3	0	0	-23	23

Table 3.

Continue by copying and using the **Paste Special** command to create the fourth matrix, and make sure you create the formulas by referencing the previously completed matrix, until the process is complete.

The Solver.

Excel's built-in **Solver** can find solutions to many different types of systems of equations, using a method similar to one you will learn in Chapter 4 of your textbook. Among the types of systems **Solver** can solve are systems of linear equations with unique solutions. The **Solver** should be located in the **Tools** menu of Excel. If it is *not*, you will need to locate your original installation disk, select **Add-Ins** from the **Tools** menu, and install the **Solver** program.

You should take care to set up a worksheet that contains as much helpful information as possible, including headings for rows, columns, and even individual cells, so that you can keep track of what all the entries represent. To complete Example 2 from your text using **Solver**, we can set up a worksheet similar to that in Table 4, with the following formulas in the indicated cells:

Cell	Contents
B3	=B1+D1+5*F1
B4	=3*B1+3*D1-F1
B5	=B1+3*D1+2*F1

x=		y=		z=	
x-y+5z=-6	0	=	-6		
3x+3y-z=10	0	=	10		
x+3y+2z=5	0	=	5		

Table 4.

From the **Tools** menu, select **Solver**. Type in the address of the cell containing the first formula, in this case, cell B3, as the "Target Cell", then check "Value of" and type in the value of the constant for the first equation, which is -6 in this example. The cells we wish to change are the cells next to the labels for x, y, and z, so type B1,D1,F1 into the box below "By Changing Cells". Our constraints here are the equations themselves, and we have already typed these in. Select **Add**, and type in the name of the cell containing the first formula. Select "=" from the drop-down menu, then type in the name of the cell containing the constant from the first equation. If your worksheet is set up like Table 4, then the **Add** window should appear as in Figure 1. If so, click **Add**.

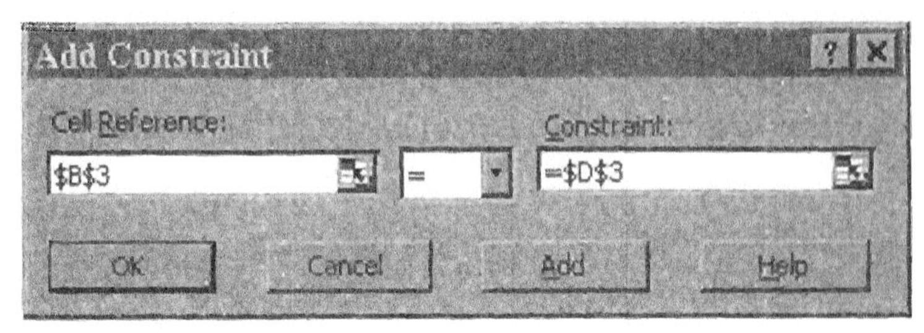

Figure 1.

Now, we need to add the second equation as a constraint, so type in B4 as the "Cell Reference", select "=" from the drop-down menu, and enter D4 as the "Constraint". We need to add one more equation, so click **Add** again and repeat this process for the third equation. When it is entered, click **OK**. This will return you to the **Solver** window, which should now look like Figure 2 on the next page.

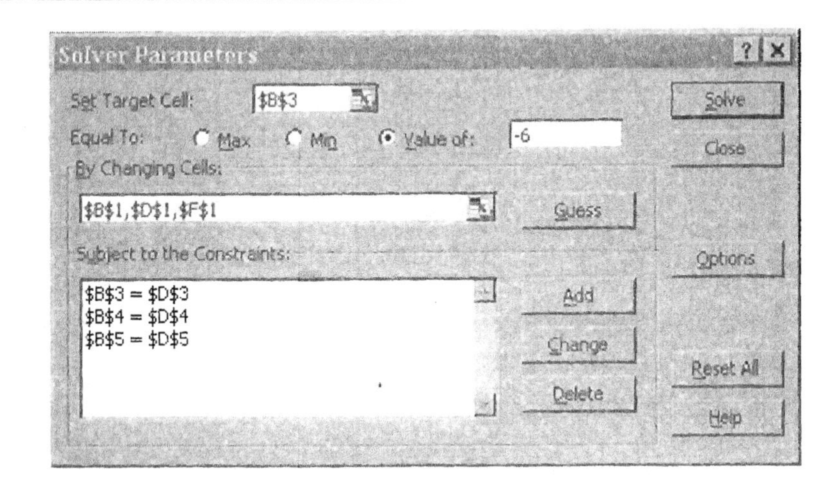

Figure 2.

Click **Solve**. The program will now adjust the values of the variables until all constraints are satisfied, if possible. The top row of the worksheet should now contain the solution to the system of equations, $x = 1$, $y = 2$, and $z = -1$.

The **Solver** cannot solve systems of equations that have no solution, nor can it solve systems with infinitely many solutions. If you try to use it to solve one of these types of systems, a new window will appear letting you know that **Solver** was unable to complete the problem.

Addition and Subtraction of Matrices.

Example 7 of this section of your text can be completed with Excel by first entering the matrices, *C* and *K*, then using a **Ctrl-Shift-Enter** trick as we did with the matrix row operations. This technique is highly useful if there are large matrices that need to be added or subtracted.

To complete Example 7, enter the matrices *C* and *K* into a worksheet, in a manner similar to Table 5. Then, highlight a block of empty cells that is the *same size as the answer should be.* In this case, you need to highlight a block of cells that is two rows by three columns. While the empty block is highlighted, type `=A2:C3-E2:G3`. This indicates that the first matrix is located in the cell range `A2:C3` and we wish to subtract from it the second matrix, whish is located in the range `E2:G3`. Press **Ctrl-Shift-Enter** and the empty block will be filled with the resulting matrix. (See Table 5 on the next page.)

C				K		
22	25	38		5	10	8
31	34	35		11	14	15
C-K						
17	15	30				
20	20	20				

Table 5.

Multiplication of Matrices.

The **MMULT** function in Excel can be used to multiply matrices. To complete Example 4 of this section of your text, enter the two matrices to be multiplied, *A* and *B*, and highlight a block of empty cells that is the same size as *AB* should be. In this case, we need to highlight a block of cells that is three rows by three columns. Type `=MMULT(A2:B4,D2:F3)` and press **Ctrl-Shift-Enter** to see the resulting product. Similarly, the product *BA* can be calculated by highlighting a set of empty cells that is 2 rows by 2 columns, typing `=MMULT(D2:F3,A2:B4)`, and pressing **Ctrl-Shift-Enter**. (See Table 6 .)

A			B		
1	-3		1	0	-1
7	2		3	1	4
-2	5				
AB				BA	
-8	-3	-13		3	-8
13	2	1		2	13
13	5	22			

Table 6.

Matrix Inverses.

Once a square matrix has been entered into a worksheet, its inverse can be calculated with the **MINVERSE** function. To complete Example 1 from this section of your text, enter matrix *A*, then highlight a set of empty cells that is the same size as the inverse should be, which is in this case 3 rows by 3 columns. Type `=MINVERSE(A2:C4)` and press **Ctrl-Shift-Enter** to calculate the inverse. (See Table 7.)

A				A-inverse		
1	0	1		0	0	0.333333
2	-2	-1		-0.5	-0.5	0.5
3	0	0		1	0	-0.33333

Table 7.

A Note About Round-off Errors and Matrix Inverses.

Occasionally, when performing calculations involving matrix inverses, one or more entries in the resulting matrix may look like `1E-10`, or something similar. Keep in mind that this represents 1 times 10 to the -10th power, or 0.0000000001. A result like this in a problem that originally contained no numbers of this form is due to round-off error during Excel's calculations. Usually, it is safe to treat any similar results as "0."

Chapter 4 Linear Programming: The Simplex Method

LOCATION IN THE OTHER TEXTS:

Finite Mathematics and Calculus with Applications: Chapter 4

Maximization Problems.

Excel's **Solver** can quickly find solutions to linear programming problems of all types, when a single solution exists. You will again need to carefully set up a worksheet so that the results of running the **Solver** are easy to interpret. Table 1 is a convenient way to set up Example 1 from this section of your text.

x1=		x2=	
Maximize	z=	0	
Subject to			
x1 + x2 <=65	0	<=	65
12x1 + 15x2 <= 900	0	<=	900

Table 1.

The following formulas were entered into the indicated cells:

Cell	Contents
C3	=25*B1+30*D1
B6	=B1+D1
B7	=12*B1+15*D1

Select **Solver** from the **Tools** menu. The "Target" cell is the one containing the expression to be maximized, which, in this case, is cell C3. Make sure that "Max" is selected here since this is a maximization problem. The cells to be changed are the cells next to the variable labels, which in Table 1 are cells B1 and D1. Add the two constraints as described in our previous discussion of **Solver**. Click **Options** and check "Assume non-negative", then click **OK**. Now click **Solve**, and the solution is calculated.

Notice that, when adding constraints, the drop-down menu allows you to use "=", "≥", and "≤". Thus, minimization problems as well as non-standard problems with single solutions, all discussed later in Chapter 4 of your text, can be completed using **Solver**. Simply choose the appropriate symbol when adding constraints into **Solver**.

Chapter 5 Mathematics of Finance

LOCATION IN THE OTHER TEXTS:

Finite Mathematics and Calculus with Applications: Chapter 5

Simple and Compound Interest.

Comparing Interest Schemes.

A spreadsheet can be used to compare two interest schemes, as shown in Figure 2 of this section of your text. To create a similar spreadsheet, enter appropriate column headings to represent years, interest compounded annually, and simple interest. (See Table 1.) Fill column A with the numbers 1 through 20 to indicate the years since the principal was invested, then type in the following:

Cell	Contents
B2	=1000*(1+0.1)^A2
C2	=1000*(1+0.1*A2)

Drag and fill columns B and C to complete the table. (Only the first 11 rows of the resulting table are shown here.)

Years	Interest Compounded Annually	Simple Interest
1	$1,100.00	$1,100.00
2	$1,210.00	$1,200.00
3	$1,331.00	$1,300.00
4	$1,464.10	$1,400.00
5	$1,610.51	$1,500.00
6	$1,771.56	$1,600.00
7	$1,948.72	$1,700.00
8	$2,143.59	$1,800.00
9	$2,357.95	$1,900.00
10	$2,593.74	$2,000.00

Table 1.

Effective Rate.

Excel has many built-in functions for calculating financial values. These are located in the **Financial** submenu of the **Function** menu. The function **EFFECT** calculates the effective interest rate corresponding to a given nominal interest and a given number of payments per year. To complete Example 6 from this section of your text, type the following in any blank cell in a worksheet:

Cell	Contents
Any cell	`=EFFECT(.06,2)`

Present Value.

The **PV** function calculates the present value of an account, when the nominal interest rate, total number of interest payments, and the future value of the account are known. In any empty cell, choose **Insert**, then **Function**, and select **PV** from the **Financial** submenu. A pop-up window will appear (see Figure 1).

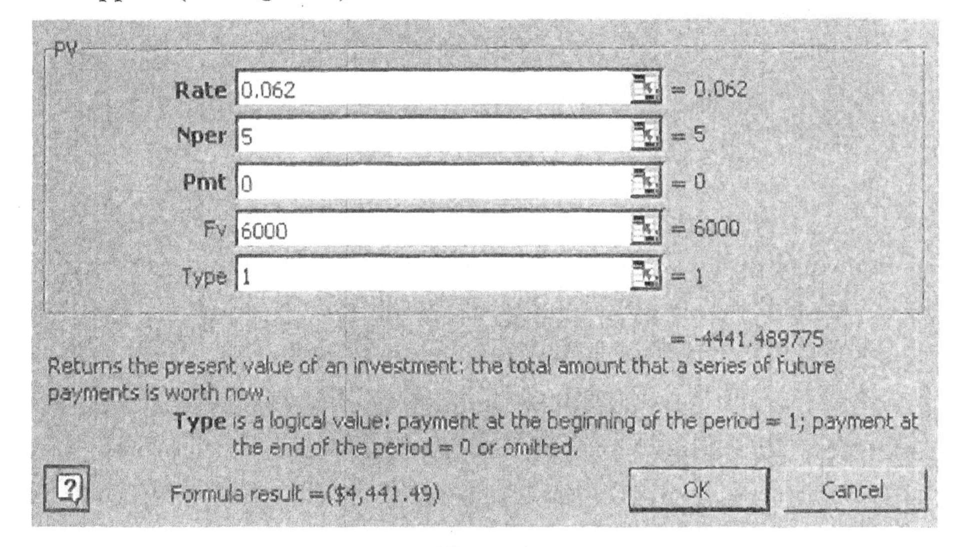

Figure 1.

Enter the interest rate, as a decimal; the total number of interest payments over the life of the account; the amount that will be paid into the account (by the account holder) at the end of each payment period; then the desired future value of the account. Figure 1 shows the appropriate entries for completing Example 9 from this section of your text. Notice that `1` was entered as the "Type" of account; this indicates that the account holder will make a payment at the beginning of a period. However, in this example, the account holder will be making only one payment, not periodic payments; this is why `0` was entered for "Pmt". When you click **OK**, the resulting present value will be entered into the selected cell. This value will be shown as a *negative* value, since this amount must be paid out by the account holder.

Future Value of an Annuity.

Ordinary Annuities.

Excel has a built-in function, **FV**, for calculating the future value of an ordinary annuity. This function is also located in the **Financial** collection of functions. From an empty cell, select **Insert**, then **Function**, then **FV** from the **Financial** submenu. Figure 2 shows the appropriate values for completing Example 4 from this section of your text.

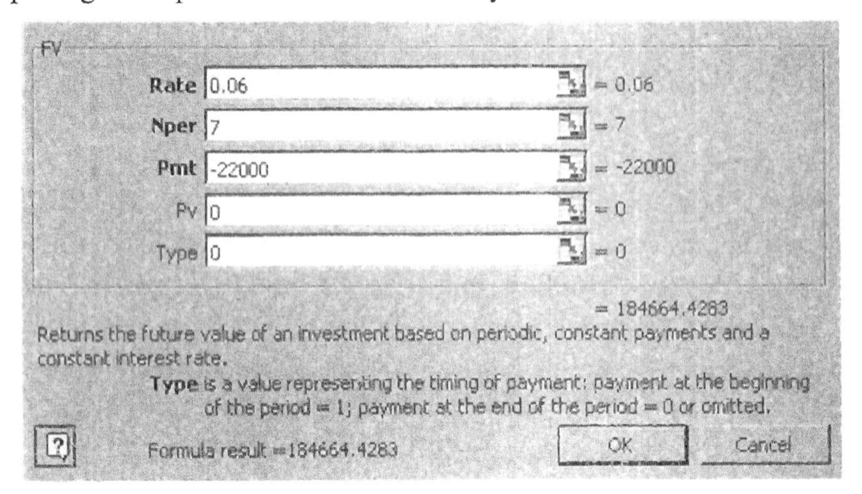

Figure 2.

Notice that, this time, 0 was selected as the "Type"; this is because Example 4 clearly states that money will be deposited at the *end* of each year. The payment, "Pmt", is entered as a *negative* number since this amount will be paid out by the account holder at the end of each payment period. Also note, that for an ordinary annuity, 0 should be entered as the present value.

If, as in Example 5, interest is compounding a certain number of times per year, this should be reflected in the "Rate" entered. For instance, in Example 5(a), 7.2% interest is compounded monthly, so we would enter 0.072/12 as the "Rate." (See Figure 3.)

FV
Rate 0.072/12 = 0.006
Nper 12*20 = 240
Pmt -200 = -200
Pv 0 = 0
Type 0 = 0
= 106752.4678
Returns the future value of an investment based on periodic, constant payments and a constant interest rate.
Rate is the interest rate per period.
Formula result = 106752.4678
OK
Cancel

Figure 3.

Present Value of an Annuity; Amortization.

Amortization Schedules.

To create an amortization schedule with Excel, similar to the one in Figure 13 from this section of your text, begin by entering appropriate column headings. (See Table 2.) Fill column `A` with the numbers `0` through `12`, indicating the payment numbers for this example. You may wish, at this point, to format the cells in columns `B`, `D`, and `E` so that numerical values are displayed as currency. (See Part I of this manual for information on formatting cells.) In cell `E2`, enter the amount of the loan, $1000. Since monthly payments are $88.85, fill cells `B3:B14` with this amount. Since interest is 1% per month (12% compounded monthly), and payments are applied to interest before principal, enter the following formulas into the indicated cells:

Cell	Contents
`C3`	`=0.01*E2`
`D3`	`=B3-C3`
`E3`	`=E2-D3`

Drag and fill the rest of columns `C`, `D` and `E`. Since the last entry is negative, the final payment should probably be adjusted by the lender to $88.83, so that the customer avoids overpaying the loan.

Payment Number	Amount of Payment	Interest for Period	Portion to Principal	Principal at End of Period
0				$1,000.00
1	$88.85	$10.00	$78.85	$921.15
2	$88.85	$9.21	$79.64	$841.51
3	$88.85	$8.42	$80.43	$761.08
4	$88.85	$7.61	$81.24	$679.84
5	$88.85	$6.80	$82.05	$597.79
6	$88.85	$5.98	$82.87	$514.91
7	$88.85	$5.15	$83.70	$431.21
8	$88.85	$4.31	$84.54	$346.67
9	$88.85	$3.47	$85.38	$261.29
10	$88.85	$2.61	$86.24	$175.05
11	$88.85	$1.75	$87.10	$87.96
12	$88.85	$0.88	$87.97	-$0.02

Table 2.

Chapter 8 Counting Principles; Further Probability Topics

LOCATION IN THE OTHER TEXTS:

Finite Mathematics and Calculus with Applications: Chapter 8

The Multiplication Principle; Permutations.

Factorial.

Excel's command for calculating factorials is the **FACT** function. To calculate 5!, type the following:

Cell	Contents
Any cell	`=FACT(5)`

Permutations.

The permutation function is **PERMUT**. To calculate P(10,3), type the following in any empty cell:

Cell	Contents
Any cell	`=PERMUT(10,3)`

Combinations.

To calculate $\binom{8}{3}$ with Excel, we use the **COMBIN** function. In any empty cell, enter the following:

Cell	Contents
Any cell	`=COMBIN(8,3)`

Detailed Instructions for Calculus with Applications

This section of Part V contains detailed instructions for using Microsoft Excel for *Calculus with Applications*, ninth edition, *Calculus with Applications: Brief Version*, ninth edition, and *Finite Mathematics and Calculus with Applications*, eighth edition. (Instructions for Chapter 1 of all three texts begin on page V-3.) The section is organized by chapters in *Calculus with Applications*; since not all chapters require detailed explanations of spreadsheet use, some chapters are not mentioned here.

In this manual, section titles from the textbooks are indicated in italics. References are made to specific examples and exercises from the corresponding sections of each chapter, so you should have your textbook nearby as you read through these instructions.

Chapter 2 Nonlinear Functions

LOCATION IN THE OTHER TEXTS:

Calculus with Applications, Brief Version: Chapter 2
Finite Mathematics and Calculus with Applications: Chapter 10

Quadratic Functions.

Quadratic Regression.

In Exercise 59 of this section of your text, a quadratic function is fit to a set of data representing the median age of women at their first marriage. The quadratic regression feature of Excel can be used to complete this exercise. There are two ways to do this, but the easiest by far is to begin by creating a scatter plot of the data set. Begin by entering appropriate column headings for the year and the median age. (See Table 1.)

Year (Since 1900)	Age
40	21.5
50	20.3
60	20.3
70	20.8
80	22.0
90	23.9
100	25.1

Table 1.

Enter the data, using the numbers 40 through 100 to represent the years 1940 through 2000. Highlight the data set, including column headings, and select **Insert**, then **Chart**. Choose the **XY (Scatter)**. Click **Finish** to see the chart. Right-click on any data point in the chart and select **Add trendline**. Select the **Polynomial** type, and make sure that "2" is selected as the "order" since we wish to fit a second degree polynomial to the data. Click the **Options** tab and select "Display equation on chart". Click **OK**. The quadratic model, and its equation, will be displayed on the chart. The result should appear similar to Figure 1 on the next page. (You may wish to drag and move the equation box to keep it from interfering with the rest of your chart.)

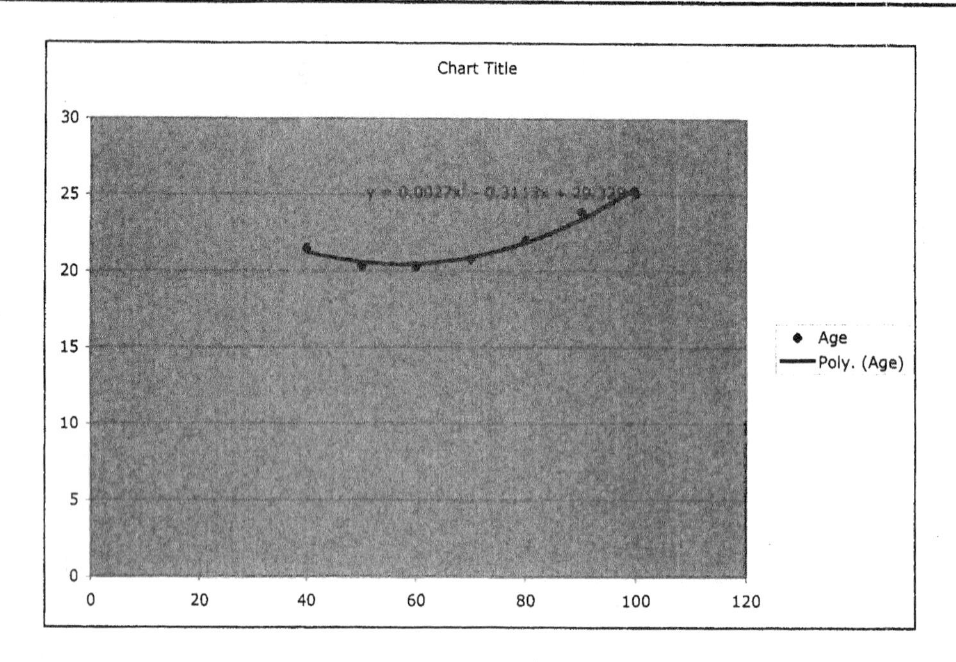

Figure 1.

This polynomial trendline feature of Excel can be used to fit up to a sixth degree polynomial model to a set of data. To do so, simply change the "order" to the degree of the polynomial you wish to fit.

Exponential Functions.

Exponential Regression.

Example 7(b) of this section of the text asks us to find an exponential function that models the given corn production data. For simplicity, subtract 1930 from each year, so that 1930 becomes year 0, 1940 becomes year 10, and so on. After entering appropriate column headings, enter the eyars and the corresponding corn production levels in adjacent columns. Follow the instructions given above for creating a scatter plot and adding a trendline. For the trendline type, select **Exponential.**

Applications: Growth and Decay; Mathematics of Finance.

Power Regression.

In Exercise 105(d) from the review exercises of your textbook, you are asked to use power regression to fit an appropriate model to a set of data. Following the same procedures as outlined above, use the **Power** type when adding a trendline.

Chapter 3 The Derivative

LOCATION IN THE OTHER TEXTS:

Calculus with Applications, Brief Version: Chapter 3
Finite Mathematics and Calculus with Applications: Chapter 11

Limits.

Creating Limit Tables.

As indicated in the discussion following Example 1, limits of functions can be estimated by observing appropriate values of a function in a table. To use Excel to estimate the limit in this example, begin by choosing appropriate column headings and entering, in column `A`, a list of x-values that surround $x = 2$. (See Table 1.) Enter the following contents into cell `B2` and drag to fill the rest of column `B`:

Cell	Contents
`B2`	`=(A2^2+4)/(A2-2)`

x	(x^2+4)/(x-2)
1.9	-76.1
1.99	-796.01
1.999	-7996.001
1.9999	-79996.0001
2.0001	80004.0001
2.001	8004.001
2.01	804.01
2.1	84.1

Table 1.

Rates of Change.

Estimating Instantaneous Rates of Change.

To generate the table in Figure 26 from this section of your text, choose appropriate column headings, like those in Table 2, for the value of h, the value of $f(x+h)$, the value of $f(x)$, and the value of the difference quotient. Enter the following values into the appropriate cells:

Cell	Contents
A2	1
B2	=3870*0.970^(20+A2)
C2	=3870*0.970^20
D2	=(B2-C2)/A2

Enter additional values of h in column A, making sure that each is smaller than the previous one, and then drag and fill columns B, C and D to create a table similar to Table 2.

h	f(x+h)	f(x)	Inst. Rate of Change
1	2041.34958	2104.48411	-63.13452321
0.1	2098.08377	2104.48411	-64.00339395
0.01	2103.8432	2104.48411	-64.09115674
0.001	2104.42001	2104.48411	-64.09994185
0.0001	2104.4777	2104.48411	-64.10082045

Table 2.

It should be clear from observing the last column of Table 2 that, as h approaches 0, the instantaneous rate of change of $f(x)$ is approximately -64.1 when $x = 20$.

Chapter 7 Integration

LOCATION IN THE OTHER TEXTS:

Calculus with Applications, Brief Version: Chapter 7
Finite Mathematics and Calculus with Applications: Chapter 15

Area and Definite Integrals.

Summation.

To use Excel to complete Example 1 from this section of your text, type in appropriate column headings, similar to those in Table 1. In column `A`, type in the numbers `1` through `4`, to indicate the number of the rectangle. In column `B`, type in the x-values representing the midpoints of the bases of the rectangles. Enter the following contents into the indicated cells:

Cell	Contents	
`C2`	`=2*B2` (evaluate $f(x)$ at `B2`)	
`F1`	`1`	(the value of Δx)

Drag and fill the rest of column `C`. Since we now wish to sum the values $f(x_i)\Delta x$, we can do this with a **SUM** command. In cell `C6`, type `=SUM(C2:C5)*$F$1`. The result should look similar to Table 1.

i	xi	f(xi)		Delta-x	1
1	0.5	1			
2	1.5	3			
3	2.5	5			
4	3.5	7			
	Sum:	16			

Table 1.

Numerical Integration.

The Trapezoidal Rule.

As indicated in the discussion after Example 1 in your text, the trapezoidal rule can be done quickly with Excel. Begin by creating column headings for the i, x_i, and $f(x_i)$. To complete

Example 1, fill column A with the integer values 0 through 4. Next, enter the following contents into the indicated cells:

Cell	Contents	
B2	=$F$1+A2*$F$2	
C2	=SQRT(B2^2+1)	(evaluate the function)
F1	0	(the left endpoint for the definite integral)
F2	0.5	(the value of Δx)

Notice the absolute cell references for F1 and F2, since you will want the same left endpoint and Δx throughout. Highlight cells B2 and C2, then drag to fill columns B and C. Finally, to apply the trapezoidal rule, in cell F3, type =F2*(0.5*C2+SUM(C3:C5)+0.5*C6). The result should appear similar to Table 2 below.

i	xi	f(xi)		Left endpoint	0
0	0	1		Delta-x	0.5
1	0.5	1.118034		Approximation	2.976529
2	1	1.414214			
3	1.5	1.802776			
4	2	2.236068			

Table 2.

Simpson's Rule.

To complete Example 2 via Simpson's rule on Excel, set up a table exactly like the one you set up for the trapezoidal rule. In cell F3, which will contain the Simpson's rule approximation for the integral, type =(F2/3)*(C2+4*C3+2*C4+4*C5+C6). The result should appear similar to Table 3 below.

i	xi	f(xi)		Left endpoint	0
0	0	1		Delta-x	0.5
1	0.5	1.118034		Approximation	2.957956
2	1	1.414214			
3	1.5	1.802776			
4	2	2.236068			

Table 3.

Chapter 10 Differential Equations

Euler's Method

Example 1 of this section of the text can be completed using Excel. Set up the sheet with columns for the independent variable x, the approximate solution variable y, the actual solution $f(x)$, and the error $y-f(x)$. You also need a single cell containing the step h. Table 1 shows the set-up with the proper initial values in the range `A1:E4`.

x	y	f(x)	y-f(x)	h
0.0	1.5	1.5	0.0	0.1

Table 1.

The contents of all the cells in the second row are numbers except for `D2`, which contains the difference formula `=B2-C2`. The third row will be all formulas, as seen in the following:

Cell	Contents
A3	`=A2+$E$2`
B3	`=B2+$E$2*(A2-2*A2*B2)`
C3	`=0.5+Exp(-(A2^2))`
D3	`=B3-C3`

Notice the absolute references for cell `E2`, as you want to use the same step size h throughout the calculations. Now select the range `A3:D3` and fill down nine rows. Your results should be identical to Table 2.

x	y	f(x)	y-f(x)	h
0	1.5	1.5	0	0.1
0.1	1.5	1.49005	0.00995	
0.2	1.48	1.460789	0.019211	
0.3	1.4408	1.413931	0.026869	
0.4	1.384352	1.352144	0.032208	
0.5	1.313604	1.278801	0.034803	
0.6	1.232243	1.197676	0.034567	
0.7	1.144374	1.112626	0.031748	
0.8	1.054162	1.027292	0.026869	
0.9	0.965496	0.944858	0.020638	
1	0.881707	0.867879	0.013827	

Table 2.